Understanding Fats & Oils

Understanding Fats & Oils

YOUR GUIDE TO HEALING WITH ESSENTIAL FATTY ACIDS

Michael T. Murray, N.D.
AND Jade Beutler, R.R.T., R.C.P.

OTHER SUGGESTED READING

Available from Progressive Health Publishing

Flax for Life!, by Jade Beutler

Fats That Can Save Your Life, by Robert Erdmann, Ph.D.

Cover design: John Odam Design Associates
Layout: Kristi Paulson Mendola
Editing: JoAnn Koppany

Library of Congress Catalog Card Number:
96-067523
ISBN 0-9645075-2-8

Understanding Fats & Oils, Your Guide to Healing with Essential Fatty Acids is meant for educational purposes only. It is not intended as medical advice. The authors and publisher recommend consulting a qualified health care professional for any condition requiring medical advice, diagnosis, or prescription.

315 First Street #U-198
Encinitas, CA 92024

CONTENTS

THE IMPORTANCE OF GOOD OILS VERSUS BAD FATS

INTRODUCTION

In this era of fat phobia and the resulting barrage of low-fat and non-fat food products lining the grocery store shelves, a recommendation to supplement an individual's daily diet with one or two tablespoons of flaxseed oil may be puzzling to many consumers. However, flaxseed oil is extremely rich in special fats designated as *essential fatty acids.* "Essential" simply means that we must consume them in our diets and that our bodies cannot manufacture them from other dietary fats or nutrients. Research suggests that a lack of essential fatty acids, ordinarily found abundantly in flaxseed oil and other unrefined polyunsaturated vegetable oils, plays a significant role in the development of such chronic degenerative diseases as heart disease, cancer, and stroke.

Many experts estimate that approximately 80 percent of the American population consumes an insufficient quantity of essential fatty acids. This dietary insufficiency presents a serious health threat to Americans. In addition to providing the body with energy, the essential fatty acids—linoleic and linolenic acid—function in our bodies as components of nerve cells, cellular membranes, and hormone-like substances known as *prostaglandins.* Prostaglandins and the essential fatty acids play an important role in keeping the body in good working order, such as

- producing steroids and synthesizing hormones
- regulating pressure in the eye, joints, or blood vessels
- regulating response to pain, inflammation, and swelling
- mediating immune response
- regulating bodily secretions and their viscosity
- dilating or constricting blood vessels

- regulating collateral circulation
- directing endocrine hormones to their target cells
- regulating smooth muscle and autonomic reflexes
- being primary constituents of cellular membranes
- regulating the rate at which cells divide (mitosis)
- maintaining the fluidity and rigidity of cellular membranes
- regulating the in-flow and out-flux of substances into and out of the cells
- transporting oxygen from red blood cells to the tissues
- maintaining proper kidney function and fluid balance
- keeping saturated fats mobile in the blood stream
- preventing blood cells from clumping together (conglomeration—the cause of atherosclerotic plaque and blood clots, a cause of stroke)
- mediating the release of pro-inflammatory substances from cells that may trigger allergic conditions
- regulating nerve transmission
- stimulating steroid production
- being the primary energy source for the heart muscle

As well as playing a critical role in normal physiology, essential fatty acids are shown to be therapeutic and protect against heart disease, cancer, autoimmune diseases such as multiple sclerosis and rheumatoid arthritis, many skin diseases, and others (see Appendix B).

Causes of Essential Fatty Acid Deficiency

The adulteration of polyunsaturated oils caused by mass commercial refinement of foods containing fats and oils has effectively eliminated the essential fatty acids from our food chain. In addition, there has been a tremendous increase in the amount of unnatural fats and oils added to the diet in the form of trans fatty acids and partially hydrogenated oils. Trans fatty acids result when polyunsaturated oils are subjected to excessive heat, light, oxygen, or other refining methods. The term *trans* literally means that the formerly C shaped (cis) polyunsaturated fatty acid is (trans)formed to an unnatural straight-shaped fatty acid molecule. Hydrogenation is caused when liquid polyunsaturated fatty acids are infused with hydrogen molecules causing an occupation of the formerly unsaturated bond with hydrogen. The result is a semisolid or solid fat substance not duplicated anywhere in nature. Margarine is the ultimate representation of a hydrogenated fat substance containing both hydrogenated and trans fatty acids.

Early in the twentieth century, Americans consumed about 125 grams of fat a day. Today, the consumption is closer to 175 grams, a 40 percent increase, or about 50 extra pounds a year. Proportionally our ingestion of saturated fats has remained relatively stable. Our ingestion of unrefined polyunsaturated oils rich in the disease-preventing essential fatty acids has decreased dramatically. Conversely, our ingestion of refined, adulterated polyunsaturated oil products has risen sharply, correlating with the dramatic rise in many degenerative conditions including cancer, heart disease, and stroke. These refined and processed compounds actually inhibit the body's ability to use the essential fatty acids that are consumed. And because synthetic fats have been prevalent in the diet for only about a hundred years, our bodies have not yet had time to evolve to the point where they can handle these deadly compounds.

There are three primary factors contributing to our current essential fatty acid deficiency.

1. Unavailability of quality oils rich in essential fatty acids because of mass commercialization and refinement of fats and oils.
2. Transformation of healthful omega-3 and omega-6 oils into toxic compounds (hydrogenated and trans isomers).
3. Metabolic competition between hydrogenated and trans fatty acids with the essential fatty acids.

Recognizing Essential Fatty Acid Deficiency

The signs and symptoms of essential fatty acid deficiency may be overt or chronically nagging, ranging from mild fatigue to a fatal heart attack. Most orthodox health care practitioners will never make the association between a health problem and essential fatty acid deficiency because they are not trained in nutrition, and the laboratory analysis to measure essential fatty acid deficiency is not widely available or appreciated. In addition, the symptoms of essential fatty acid deficiency are not as obvious as with many other nutrient deficiencies. The consequences of this lack of knowledge can be deadly. And even if an essential fatty acid deficiency were recognized, few orthodox clinicians would know how to treat it.

The symptoms of essential fatty acid deficiency can be so vague and broad that they are usually written off as having some other cause. Surveys suggest that most Americans are obtaining only about 10 percent of what they need for optimal health. This is why the authors believe that everyone, regardless of health status, should take essential fatty acid–rich flaxseed oil. The following guidelines should help you recognize your personal essential fatty acid (EFA) status.

Signs and symptoms typical of, but not exclusive to, EFA deficiency:

- aching joints
- angina; chest pain
- arthritis
- constipation
- cracked nails
- depression
- dry, lifeless hair
- dry mucous membranes, tear ducts, mouth, vagina
- dry skin
- fatigue; malaise; lackluster energy
- forgetfulness
- frequent colds and sickness
- high blood pressure
- history of cardiovascular disease
- immune weakness
- indigestion; gas; bloating
- lack of endurance
- lack of motivation

Some Practical Advice

Here are four recommendations to achieve better health and more optimal levels of essential fatty acids in body tissues.

1. *Reduce the amount of saturated fats and total fat in the diet.* There is much research linking saturated fats to numerous cancers, heart disease, and strokes. Both the American Cancer Society and the American Heart Association have recommended a diet containing less than 30 percent of calories as fat. Looking at the chart on page 5, it is obvious that the easiest way for most people to achieve this goal is to eat less animal products and more plant foods. With the exception of nuts and seeds, most plant foods are very low in fat. And though nuts and seeds do contain high levels of fat calories, the calories are derived largely from polyunsaturated essential fatty acids.
2. *Eliminate the intake of margarine and other foods containing trans fatty acids and partially hydrogenated oils.* During the manufacturing process of margarine and shortening, vegetable oils are hydrogenated; that is, a hydrogen molecule is added to the natural unsaturated fatty acid molecules of the vegetable oil to make it more saturated. This change in structure of the natural fatty acid to many "unnatural" fatty acid forms interferes with the body's ability to utilize essential fatty acids.
3. *Take one or two tablespoons of flaxseed oil daily.* Organic, unrefined flaxseed oil is considered by many to be the answer to restoring the proper level of essential fatty acids. Flaxseed oil is unique because it contains both essential fatty acids—alpha linolenic (omega-3) and linoleic (omega-6)—in appreciable amounts. Flaxseed oil is the richest source of omega-3 fatty acids. At a whopping 58 percent by weight, it contains over twice the

amount of omega-3 fatty acids as fish oils. Omega-3 fatty acids have been extensively studied for their beneficial effects on high cholesterol levels, high blood pressure, stroke and heart attack, angina, arthritis, multiple sclerosis, inflammatory skin disorders, and inhibiting cancer formation and metastasis.

PERCENTAGE OF CALORIES AS FAT

Eggs & Dairy Products

Food	%
Butter	100%
Cream, light whipping	92%
Cream cheese	90%
Egg yolks	80%
Half-and-half	79%
Cheddar cheese	71%
Swiss cheese	66%
Eggs, whole	65%
Cow milk	49%
Yogurt, plain	49%
Ice cream, regular	48%
Cottage cheese	35%
Lowfat (2%) milk/yogurt	31%

Meats

Food	%
Sirloin steak*	83%
Pork sausage	83%
T-bone steak*	82%
Porterhouse steak*	82%
Bologna	81%
Spareribs	80%
Frankfurters	80%
Lamb rib chops*	79%
Salami	76%
Rump roast*	71%
Ham*	69%
Ground beef, fairly lean	64%
Veal breast*	64%
Leg of lamb	61%
Round steak*	61%
Chicken, dark meat+	56%
Chuck steak, lean only	50%
Turkey, dark meat w/skin	47%
Chicken, light meat+	44%

Fruits

Food	%
Grapes	11%
Strawberries	11%
Apples	8%
Blueberries	7%
Lemons	7%
Pears	5%
Apricots	4%
Oranges	4%
Bananas	4%
Cantaloupe	3%
Pineapple	3%
Grapefruit	2%
Papayas	2%
Peaches	2%
Prunes	1%

Vegetables

Food	%
Mustard greens	13%
Kale	13%
Beet greens	12%
Lettuce	12%
Turnip greens	11%
Cabbage	7%
Cauliflower	7%
Green beans	6%
Celery	6%
Cucumbers	6%
Turnips	6%
Zucchini	6%
Carrots	4%
Green peas	4%
Beets	2%
Potatoes	1%

* Lean, with fat + With skin, roasted

Source: "Nutritive Value of American Foods in Common Units," U.S.D.A. Agriculture Handbook No. 456

4. *Limit total dietary fat intake to no more than 30 percent of calories consumed* (400–600 calories a day, based on a standard 2000-calorie-a-day diet). Make a strong effort to incorporate "healthful" fats in the form of essential fatty acid–rich oils such as flaxseed oil in

place of dangerous trans, hydrogenated, and saturated fats. Watch for these "stealth fats" by reading food labels carefully before you choose.

The authors hope this brief overview has inspired you to learn more about the role of essential fatty acids in health and disease.

UNDERSTANDING THE TERMINOLOGY

1

In order to understand the harmful effects of some fats and the beneficial effects of others, it is important to understand the terminology used to describe fats and oils. Let's start with the meaning of "fat." *Fat,* or *lipid,* refers to a number of compounds composed of carbon, hydrogen, and oxygen that are not soluble in water. The three major classes of dietary fats are triglycerides, phospholipids, and sterols (like cholesterol). We will examine each of these in subsequent chapters.

Triglycerides

The most common chemical form of fats in the diet are *triglycerides.* They comprise approximately 95 percent of all ingested fats. A triglyceride is composed of a *glycerol* molecule and three fat molecules known as *fatty acids* (see Figure 1.1). A fatty acid is composed of a long chain of carbon molecules with an acid group on one end.

The body handles dietary triglycerides by breaking the bond between glycerol and the fatty acids. This breaking of the chemical bond is achieved by first emulsifying the triglyceride with bile so that enzymes, known as *lipases,* can add water to the glycerol molecule, thus liberating a fatty acid. Initially, the triglyceride is converted into a diglyceride, then a monoglyceride. The body can absorb free fatty acids and glycerol as well as monoglycerides much easier than the bulkier triglycerides and diglycerides (see Figure 1.2).

Once digested, the free fatty acids and monoglycerides are absorbed into the body and transported by special protein-wrapped molecules known as *lipoproteins.* The major categories of lipoproteins are very low-density lipoprotein (VLDL), low-density lipoprotein (LDL), and high-density lipoprotein (HDL). Because VLDL and LDL are responsible for

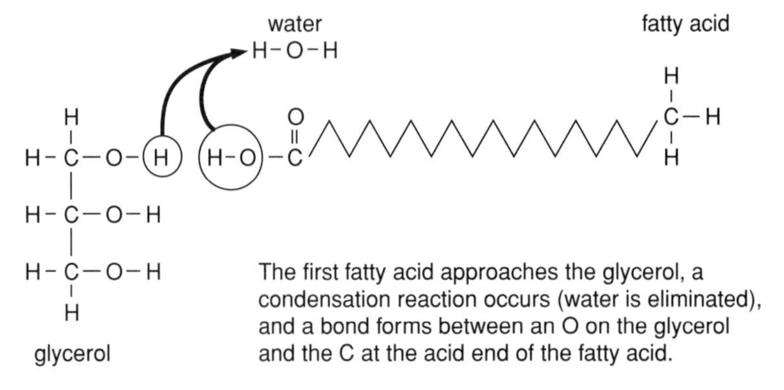

Later, two more fatty acids attach themselves to the glycerol by the same means; the resulting structure is a triglyceride.

A fat (triglyceride) that might be found in butter.

FIGURE 1.1 *Glycerol + fatty acid*

transporting fats—primarily triglycerides and cholesterol—from the liver to body cells and HDL is responsible for returning fats to the liver, elevations of either VLDL or LDL are associated with an increased risk of developing atherosclerosis, the primary cause of heart attacks and strokes. Conversely, elevations of HDL are associated with a low risk of heart attacks. The role of lipoproteins is discussed further in Chapter 4.

Saturated and Unsaturated Fats

The triglyceride shown in Figure 1.1 is a *saturated* fat because all of the carbon molecules on the fatty acids are saturated with as many hydrogen molecules as they can carry. If some of the hydrogen molecules were removed, the triglyceride would then be an *unsaturated* fatty acid.

Saturated fats are semisolid to solid at room temperature and are typically animal fats like butter, lard, and tallow. Unsaturated fats are

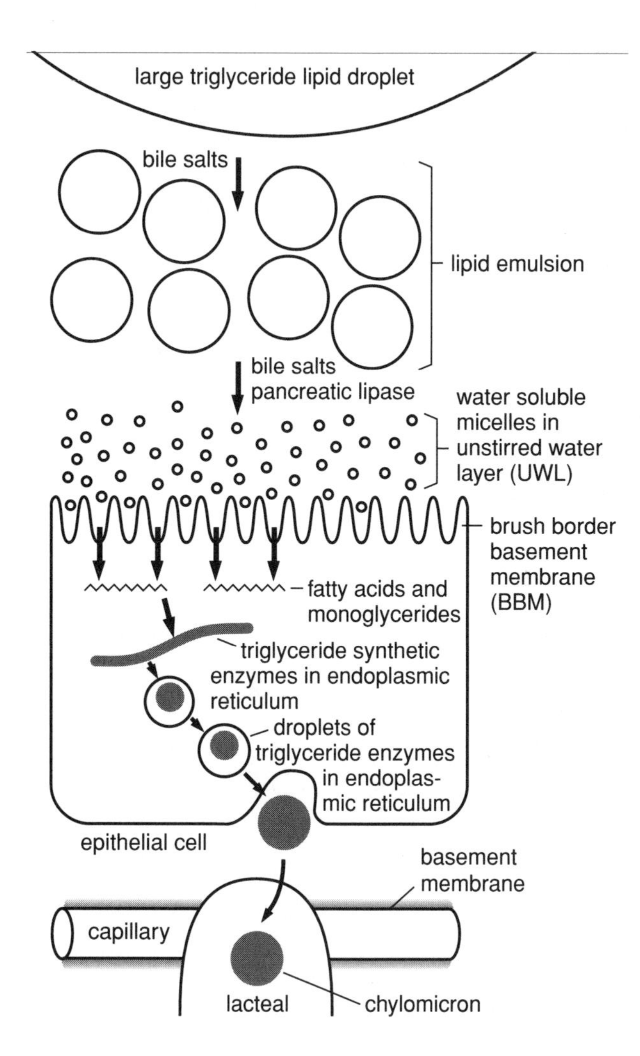

FIGURE 1.2 *Large dietary triglycerides are broken down into absorbable units (fatty acids and monoglycerides) with the aid of bile salts and enzymes (lipases) secreted by the pancreas. Once broken down into smaller units, called* micelles, *the fatty acids and monoglycerides are absorbed into the intestinal cells where they are re-formed into triglycerides and pushed out of the cell into the lymphatic system, where they will be transported to the liver for processing.*

typically liquid at room temperature and are therefore often referred to as oils. Most vegetable oils contain primarily unsaturated fats. To illustrate the difference in structure, let's look at the 18-carbon family of fatty acids.

Stearic acid is an 18-carbon-long saturated fatty acid (see Figure 1.3), which means that it is carrying as many saturated carbon molecules as it can.

Oleic acid, an 18-carbon-long monounsaturated fatty acid (see Figure 1.4), is missing two hydrogen molecules. This leaves two carbon molecules unsaturated (see Figure 1.5), causing them to bind to each other to form a double bond. Oleic acid is monounsaturated and is termed an omega-9 oil because its first unsaturated bond occurs at the ninth carbon from the omega end (see Figure 1.6). To illustrate this, oleic acid is written C18:1w9, meaning that it is a carbon chain containing 18 carbon molecules and one double bond at the ninth carbon.

Linoleic acid, written C18:2w6, is an 18-carbon-long polyunsaturated fatty acid because it contains two double bonds. It is classified as an omega-6 oil because the first double bond occurs at the sixth carbon from the omega end (see Figure 1.7).

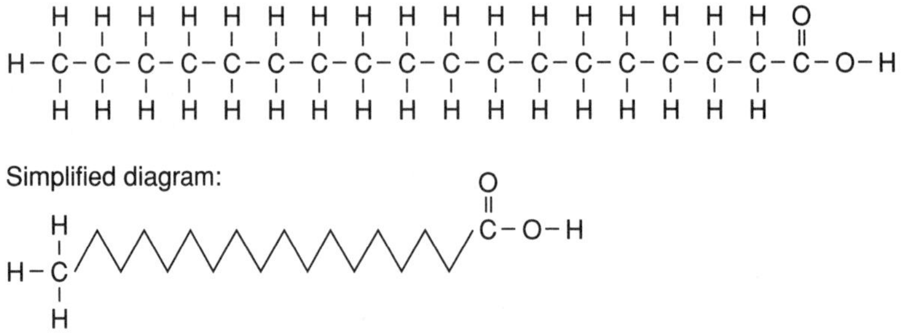

```
    H H H H H H H H H H H H H H H H H O
    | | | | | | | | | | | | | | | | | ||
  H-C-C-C-C-C-C-C-C-C-C-C-C-C-C-C-C-C-C-O-H
    | | | | | | | | | | | | | | | | |
    H H H H H H H H H H H H H H H H H

Simplified diagram:
                                O
    H                           ||
    |  /\/\/\/\/\/\/\/\/\/\/\/\/C-O-H
  H-C/
    |
    H
```

FIGURE 1.3 *Stearic acid*

```
    H H H H H H H H     H H H H H H H O
    | | | | | | | | | | | | | | | | | ||
  H-C-C-C-C-C-C-C-C-C-C-C-C-C-C-C-C-C-C-O-H
    | | | | | | | | | | | | | | | | |
    H H H H H H H H H H H H H H H H H
```

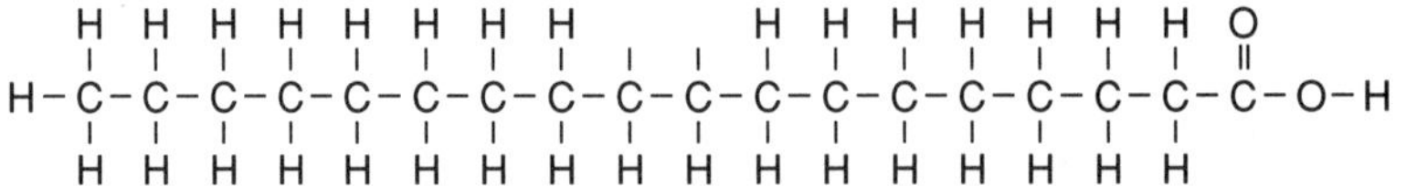

FIGURE 1.4 *Monounsaturated fatty acid*

```
    H H H H H H H H     H H H H H H H O
    | | | | | | | |     | | | | | | | ||
  H-C-C-C-C-C-C-C-C-C=C-C-C-C-C-C-C-C-C-O-H
    | | | | | | | | | | | | | | | | |
    H H H H H H H H H H H H H H H H H

Simplified diagram:
    H                              O
    |  /\/\/\/\____/\/\/\/\        ||
  H-C/                     \       C-O-H
    |
    H
```

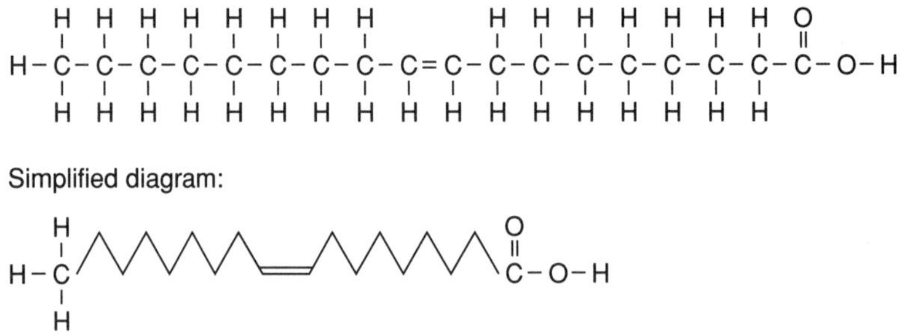

FIGURE 1.5 *Oleic acid*

Alpha-linolenic acid is an 18-carbon-long polyunsaturated fatty acid with three double bonds, C18:3w3. Alpha-linolenic acid is an omega-3 oil because its first double bond is at the third carbon (see Figure 1.8).

Fatty Acid Composition of Vegetable Oils

Vegetable oils can be loosely divided into two categories: cooking oils and medicinal oils. The differentiating factor is that the medicinal oils

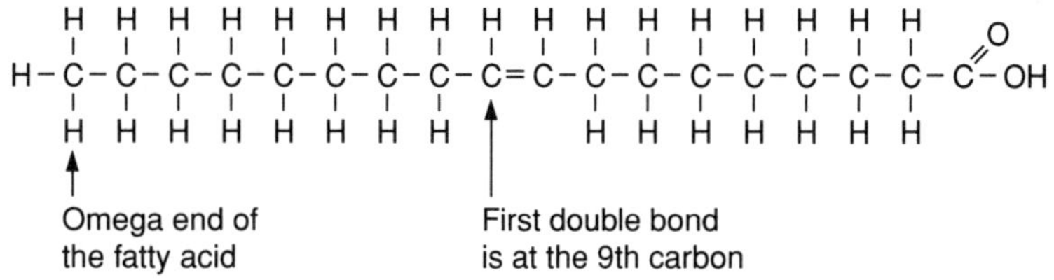

Omega end of the fatty acid

First double bond is at the 9th carbon

FIGURE 1.6 *An omega-9 oil (oleic acid)*

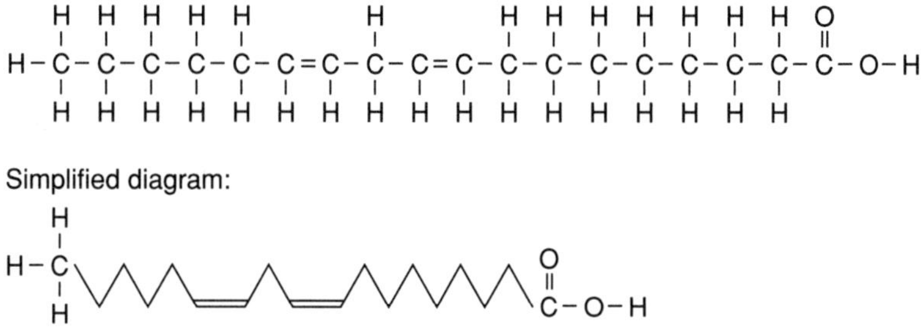

Simplified diagram:

FIGURE 1.7 *An omega-6 oil (linoleic acid)*

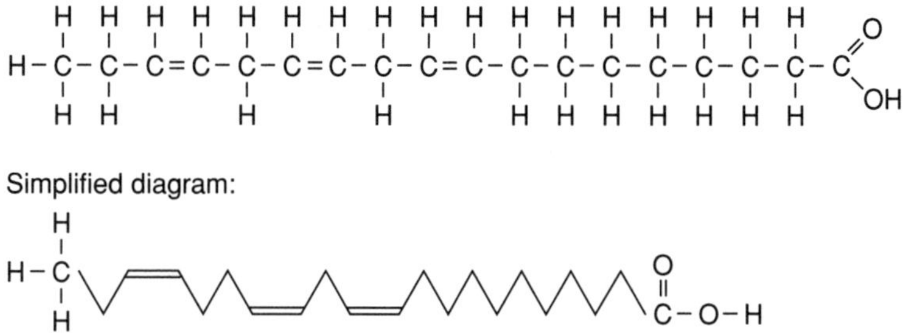

Simplified diagram:

FIGURE 1.8 *An omega-3 oil (linolenic acid)*

contain gamma-linolenic acid (evening primrose, borage, and black currant) or alpha-linolenic acid (flaxseed). These oils are also highly polyunsaturated, which means they do not hold up well when exposed to heat.

Most experts agree that the best oils for cooking are canola and olive oil because these oils are composed primarily of oleic acid, a monounsaturated oil that is more resistant to the damaging effects of heat and light compared to such highly polyunsaturated oils as corn, safflower, and soy. When the polyunsaturated oils are exposed to heat or light, the chemical structure of the essential fatty acids is changed to toxic derivatives known as *lipid peroxides* (Booyens and Van Der Merwe 1992;

Longnecker 1993; Mensink and Katan 1990; Willett et al. 1993).

As for the best vegetable oil for medicinal purposes, the authors recommend flaxseed oil because it is nature's richest source of omega-3 oils.

FATTY ACID COMPOSITION (% OF TOTAL FAT) OF SELECTED OILS

	SF	OA	LA	GLA	Alpha-LA
Cooking oils					
Canola	7	54	30	0	7
Olive	16	76	8	0	0
Soy	15	26	50	0	9
Corn	17	24	59	0	0
Safflower	7	10	80	0	0
Medicinal oils					
Evening Primrose	10	9	72	9	0
Black Currant	7	9	47	17	13
Borage	14	16	35	22	0
Flaxseed	9	19	14	0	58

SF = Saturated Fats
OA = Oleic Acid
LA = Linolenic Acid
GLA = Gamma-Linolenic Acid (omega-6 oil)
Alpha-LA = Alpha-Linolenic Acid (omega-3 oil)

What About Margarine? During the manufacture of margarine and shortening, vegetable oils are *hydrogenated.* This means that a hydrogen molecule is added to the natural unsaturated fatty acid molecules of the vegetable oil to make it more saturated. This results in the natural structure of the fatty acid to be changed to an unnatural form, causing the vegetable oil to become solid or semisolid. The configuration of the fatty acid has changed from *cis* to *trans* (see Figure 1.9).

Trans fatty acids and hydrogenated oils have been linked to low birth weight in infants, low quality and volume of breast milk, abnormal sperm production and decreased testosterone in men, heart disease, increased levels of harmful cholesterol in humans, prostate disease, obesity, suppression of the immune system, and essential fatty acid deficiencies (Enig 1993).

Many researchers and nutritionists have been concerned about the health effects of margarine ever since it was first introduced. Although many people assume they are doing their body good by consuming margarine instead of butter and other saturated fats, in truth they are actually doing more harm. Margarine and other hydrogenated vegetable oils not only raise LDL cholesterol, they also lower the protective HDL cholesterol level, interfere with essential fatty acid metabolism, and are suspected of being the cause of certain cancers (Enig 1993).

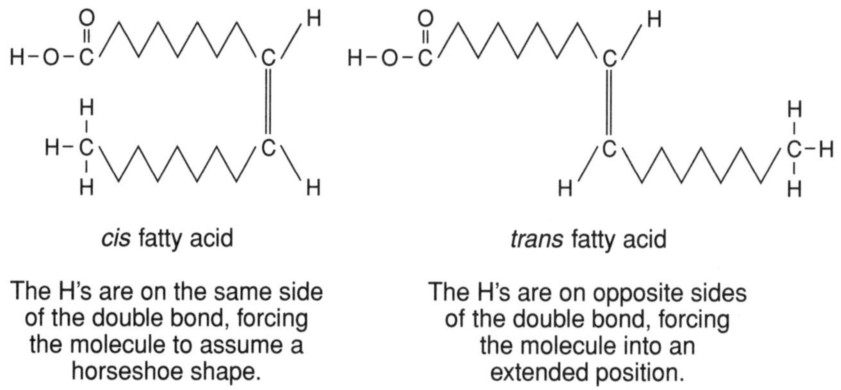

cis fatty acid

The H's are on the same side of the double bond, forcing the molecule to assume a horseshoe shape.

trans fatty acid

The H's are on opposite sides of the double bond, forcing the molecule into an extended position.

FIGURE 1.9 Cis *vs.* trans *fatty acid configuration*

Although butter may be better than margarine, the bottom line is that they both need to be restricted in a healthy diet while natural polyunsaturated oils like canola, safflower, soy, and flaxseed should be used to meet essential fatty acid requirements: Just one tablespoon of a high-quality flaxseed oil will provide more than enough in most cases.

Foods typically containing partially or totally hydrogenated vegetable oils and trans isomers:

all refined and processed foods
bread
cakes
candies
canned soups and food
cereals
cookies
crackers
doughnuts
margarine
processed cheese
snack foods
grocery store oils: canola, corn, safflower, sesame, sunflower, walnut

References

Booyens, J., and Van Der Merwe, C. F. (1992). "Margarines and coronary artery disease." *Med Hypothesis* 37:241–244.

Enig, M. G. (1993). "Trans fatty acids: An update." *Nutrition Quarterly* 17(4):79–95.

Longnecker, M. P. (1993). "Do trans fatty acids in margarine and other foods increase the risk of coronary heart disease?" *Epidemiology* 4:492–495.

Mensink, R. P., and Katan, M. B. (1990). "Effect of dietary trans fatty acids on high-density and low-density lipoprotein cholesterol levels in healthy subjects." *New Engl J Med* 323:439–445.

Vijai, K.S., Perkins, S., and Perkins E. G. (1991). "The presence of oxidative polymeric materials in encapsulated fish oils." *Lipids* 26:1.
Willett, W. C., et al. (1993). "Intake of trans fatty acids and risk of coronary heart disease among women." *Lancet* 341:581–585.

WHY SATURATED FATS ARE "BAD" AND ESSENTIAL FATTY ACIDS ARE "GOOD"

2

What makes saturated fats and margarine "bad" and essential fatty acids "good" is the function of essential fatty acids in the body and the interference of this function by saturated fats and hydrogenated oils. To keep it simple, let's examine the role of essential fatty acids in cellular membranes.

The Role of Essential Fatty Acids

All cells throughout the human body are enveloped by membrane composed mainly of essential fatty acids called *phospholipids*. A phospholipid differs from a triglyceride in that, instead of three fatty acids being attached to the glycerol molecule, one of the fatty acids is replaced with a phosphorus-containing molecule like choline or serine. Most of the phospholipids in the cell membranes are manufactured by adding the phosphate group to a diglyceride.

Phospholipids play a major role in determining the integrity and fluidity of cell membranes. What determines the type of phospholipid in the cell membrane is the type of fat consumed. Although we can ingest preformed phospholipids like lecithin or phosphatidylcholine, most of these phospholipids are broken down into glycerol, free fatty acids, and the phosphate group rather than being incorporated intact into cellular membranes.

A phospholipid composed of a saturated fat or trans fatty acid differs considerably in structure from a phospholipid composed of an essential fatty acid. In addition, there are differences between the structure of an omega-3 oil composed membrane and an omega-6 oil composed membrane. The degree of difference is illustrated by the difference in fluidity among the various fatty acids.

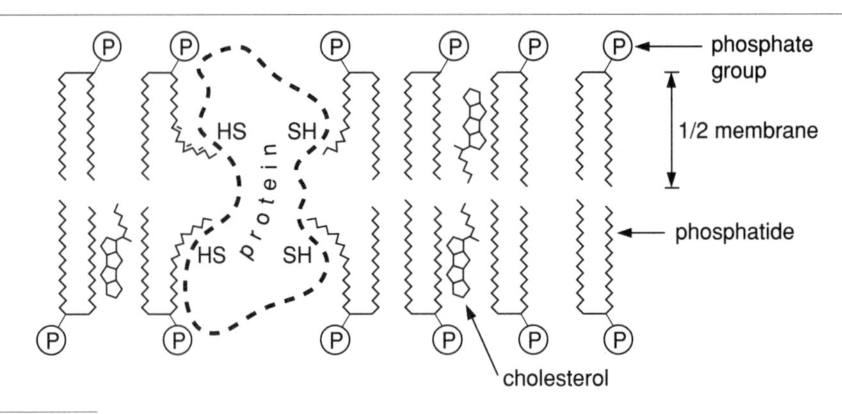

FIGURE 2.1 *Structure of a biological membrane*

Basically, it is thought the cell is programmed to selectively incorporate the different fatty acids it needs to maintain optimal function. But because of the lack of essential fatty acids (particularly the omega-3 oils) in the standard American diet, in actuality, what becomes incorporated into the cell membranes is determined primarily by diet. A diet composed largely of saturated fat, animal fatty acids (e.g., arachidonic acid), cholesterol, and trans fatty acids is going to lead to membranes that are much less fluid in nature than the membranes of an individual consuming optimum levels of omega-3 and omega-6 essential fatty acids.

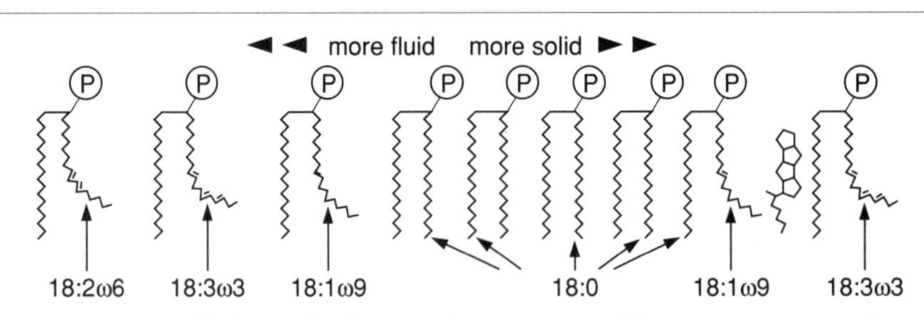

FIGURE 2.2 *How a membrane becomes more fluid or more solid*

A relative deficiency of essential fatty acids in cellular membranes makes it virtually impossible for the cell membrane to perform its vital function. The basic function of the cell membrane is to serve as a selective barrier that regulates the passage of certain materials in and out of the cell. When there is a disturbance of structure or function of the cell membrane, there is a tremendous disruption of homeostasis. Homeostasis refers to the maintenance of static, or constant, conditions in the internal environment of the cell and, on a larger scale, the human body as a whole. In other words, with a disturbance in cellular membrane structure or function, there is disruption of virtually all cellular processes.

Cell Membrane Alterations and Disease

According to modern pathology, an alteration in cell membrane function is the central factor in the development of cell injury and death (Robbins, Cotran, and Kumar 1984). Without a healthy membrane, cells lose their ability to hold water, vital nutrients, and electrolytes. They also lose their ability to communicate with other cells and be controlled by regulating hormones. They simply do not function properly. As an example, let's take a look at the effect of membrane fluidity upon the action of the hormone insulin.

Insulin's role is to stimulate the uptake of blood glucose into cells. If there is insufficient insulin or the cell does not respond to insulin, blood glucose levels can be elevated—a condition known as *diabetes mellitus*. Diabetes is divided into two major categories: type I and type II. Type I, or Insulin-Dependent Diabetes Mellitus (IDDM), occurs most often in children and adolescents. It is associated with complete destruction of the beta cells of the pancreas that manufacture insulin. Type I diabetics will require insulin for the rest of their lives for the control of blood sugar levels. Only about 10 percent of all diabetics are type I; the rest are type II.

Type II, or Non-Insulin Dependent Diabetes Mellitus (NIDDM), usually occurs after age 40. Insulin levels are typically elevated, indicating a loss of sensitivity to insulin by the cells of the body. Obesity and the type of dietary fat consumed are major contributing factors to this loss of insulin sensitivity.

The type of dietary fat profile linked to type II diabetes is an abundance of saturated fat and a relative insufficiency of essential fatty acids (National Research Council 1989; Pelikanova et al. 1991). One of the key reasons appears to be the fact that such a dietary pattern leads to reduced membrane fluidity, which in turn causes reduced insulin binding to receptors on cellular membranes or reduced insulin action (Borkman et al. 1993; Pelikanova 1989).

Conversely, omega-3 oils appear to improve insulin action (Borkman et al. 1993; Pelikanova 1989). Population studies have shown that frequent consumption of a small amount of omega-3 oils protects against the development of type II diabetes (Feskens, Bowles, and Kromhout 1991). In addition, animal studies have also shown that omega-3 fatty acids prevent the development of insulin resistance (Storlien et al. 1991). All of this evidence appears to indicate that altered membrane fluidity may play a critical role in the development of type II diabetes.

To better determine the role of specific fatty acids in increasing the risk of developing NIDDM, researchers recently examined the fatty acid composition of the serum cholesterol esters among 50-year-old men in a ten-year follow-up to the famous Uppsala study. The fatty acid composition of the serum cholesterol esters reflects the average quality of fat consumed over several weeks, or perhaps even longer periods of time. Of the 1,828 men who did not have diabetes in 1970–1973, 75 developed NIDDM (Vessby et al. 1994).

The results of the study showed striking differences in the serum cholesterol esters. The subjects with diabetes had higher proportions of saturated fatty acids and palmitoleic acid (16:1w7), a low proportion of linoleic acid (18:2w6), and relatively high (although numerically small) increases of gamma-linolenic (GLA, 18:3w6) and dihomo-gamma-linolenic acid (DHGLA, 20:3w6). The development of NIDDM was significantly predicted by a high proportion of DHGLA. The altered fatty acid profiles preceded the development of NIDDM.

The dietary factors responsible for such a fatty acid pattern are high intakes of meat and dairy products along with low intakes of vegetable oils. A high proportion of DHGLA, a significant marker for the likelihood of developing NIDDM, can be explained by an increased consumption of arachidonic acid or, possibly, by genetic factors. DHGLA is converted into arachidonic acid by an enzyme (delta-5 desaturase). However, when arachidonic acid levels are high, the activity of this enzyme is reduced via feedback inhibition.

The results of this study indicate the following dietary recommendations related to fatty acid intake might significantly reduce the risk of developing NIDDM:

1. Reduce the intake of saturated fatty acids.
2. Increase the consumption of essential fatty acids (linoleic and alpha-linolenic acids).
3. Increase consumption of omega-3 oils by consuming cold-water fish and/or flaxseed oil.

These recommendations may not only improve insulin sensitivity but also improve serum cholesterol and triglyceride levels as well.

Determinants of Healthy Cellular Membranes

Although scientists have determined that the health of the cell is critically dependent on the health and integrity of the cellular membrane, for some reason they have been slower to fully clarify the factors that determine the health of the cellular membrane itself. The research that does exist on this subject points to two important factors: adequacy of essential fatty acid intake, including alpha-linolenic acid; and adequate levels of antioxidants. Now that you understand the role of essential fatty acids, we will define and examine the role of dietary antioxidants.

Antioxidants Protect Cellular Membranes

The cell membranes of the human body are constantly under attack by free radicals and pro-oxidants. These highly reactive molecules can bind to and destroy cellular membranes as well as other cellular components. A *free radical* is a molecule that contains a highly reactive, unpaired electron, and a *pro-oxidant* is a molecule that can promote oxidative damage.

Most of the free radicals zipping through our bodies are actually produced during such normal metabolic processes as energy production, detoxification reactions, and immune defense mechanisms. In fact, the major source of free radical damage in the body is the oxygen molecule.

It is ironic that oxygen, the molecule that gives us life, is also the molecule that can do the most harm. Just as oxygen can rust iron, when toxic oxygen molecules are allowed to attack our cells, free radical or oxidative damage occurs. Our cells protect against free radical and oxidative damage with the help of enzymes and their ability to incorporate valuable antioxidant compounds from the diet such as beta-carotene, vitamins C and E, and sulfur-containing amino acids.

Although the body's own generation of free radicals is important, the environment contributes greatly to the free radical "load" of an individual. Cigarette smoking, for example, greatly increases an individual's free radical load. Many of the harmful effects of smoking are related to the extremely high levels of free radicals being inhaled, depleting key antioxidant nutrients like vitamin C and beta-carotene. Other external sources of free radicals include ionizing radiation, chemotherapeutic drugs, air pollutants, pesticides, anesthetics, aromatic hydrocarbons, fried food, solvents, alcohol, and formaldehyde. These compounds greatly stress the body's antioxidant mechanisms. Individuals exposed to these compounds will require higher antioxidant intakes.

More and more research demonstrates that a high intake of antioxidant nutrients can help prevent some of the major degenerative diseases of our society as well as possibly slow down the aging process (Diplock 1991; Hennekens and Gaziano 1993; Stahelin et al. 1991). A high antioxidant intake is also recommended for individuals consuming higher quantities of essential fatty acids. Consuming a diet rich in fresh fruits and vegetables is the first step in achieving higher antioxidant levels. The second step is taking extra antioxidant nutrients. The following supplements are recommended:

Vitamin C—1,000 to 3,000 mg daily
Vitamin E—400 to 800 IU daily
Selenium—200 mcg daily

Other Reasons Why Essential Fatty Acids are "Good"

In addition to their role in cellular membranes, essential fatty acids play critical roles in other body functions and structures. Especially important are the ways in which essential fatty acids are transformed into regulatory compounds known as prostaglandins. Once again this function of essential fatty acids is altered with increased intake of saturated fat and trans fatty acids. The importance of the omega-6 to omega-3 ratio will be discussed in Chapter 3.

References

Borkman, M., et al. (1993). "The relationship between insulin sensitivity and the fatty acid composition of skeletal-muscle phospholipids." *New Engl J Med* 328:238–244.

Diplock, A. T. (1991). "Antioxidant nutrients and disease prevention: an overview." *Am J Clin Nutr* 53:189S–193S.

Feskens, E. J. M., Bowles, C. H., and Kromhout, D. (1991). "Inverse association between fish intake and risk of glucose intolerance in normoglycemic elderly men and women." *Diabetes Care* 14:935–941.

Hennekens, C. H., and Gaziano, J. M. (1993). "Antioxidants and heart disease: Epidemiology and clinical evidence." *Clin Cardiol* 16 (Suppl.1):10–15.

National Research Council. (1989). *Diet and Health. Implications for Reducing Chronic Disease Risk.* National Academy Press, Washington, DC.

Robbins, S. L., Cotran, R. S., and Kumar, V. (1984). *Pathologic Basis of Disease,* 3rd Ed. W. B. Saunders, Philadelphia, PA.

Pelikanova, T., et al. (1989). "Insulin secretion and insulin action are related to the serum phospholipid fatty acid pattern in healthy men." *Metab Clin Exp* 38:188–192.

Pelikanova, T., et al. (1991). "Fatty acid composition of serum lipids and erythrocyte membranes in type 2 (non-insulin-dependent diabetic men)." *Metab Clin Exp* 40:175–180.

Stahelin, H. B., et al. (1991). "Plasma antioxidant vitamins and subsequent cancer mortality in the 12-year follow-up of the prospective Basel Study." *Am J Epidemiology* 133:766–775.

Storlien, L. H., et al. (1991). "Influence of dietary fat composition on the development of insulin resistance in rats: Relation to muscle triglyceride and omega-3 fatty acids in muscle phospholipid." *Diabetes* 40:280–289.

Vessby, B., et al. (1994). "The risk to develop NIDDM is related to the fatty acid composition of the serum cholesterol esters." *Diabetes* 43:1353–1357.

THE IMPORTANCE OF THE OMEGA-6 TO OMEGA-3 RATIO

3

The balance of omega-6 to omega-3 oils is critical to proper prostaglandin metabolism. Prostaglandins and related compounds are hormone-like molecules derived from 20-carbon-chain fatty acids that contain three, four, or five double bonds. Linoleic and linolenic can be converted to prostaglandins through adding two carbon molecules and removing hydrogen molecules (if necessary). Prostaglandins are important for the regulation of inflammation, pain, and swelling; blood pressure; heart function; gastrointestinal function and secretions; kidney function and fluid balance; blood clotting and platelet aggregation; allergic response; nerve transmission; and steroid production and hormone synthesis.

Prostaglandins

The number of double bonds in the fatty acid determines the classification of the prostaglandin. In humans the omega-3 and omega-6 essential fatty acids are ultimately converted via metabolic pathways to three different series of prostaglandins, each serving specific and important functions. Series 1 and 2 come from the omega-6 fatty acids with linoleic acid serving as the starting point. Linoleic acid is changed to gamma-linolenic acid and then to dihomo-gamma-linolenic acid, which contains three double bonds and is the precursor to prostaglandin of the 1 series. Dihomo-gamma-linolenic acid (DHGLA) can also be converted to arachidonic acid, which contains four double bonds and is a precursor to the 2 series prostaglandins. However, because the enzyme (delta-5 desaturase) responsible for the conversion of DHGLA to arachidonic acid prefers the omega-3 oils, in humans the greatest source of arachidonic acid comes from the diet. Arachidonic acid is found almost entirely in animal foods along with saturated fats.

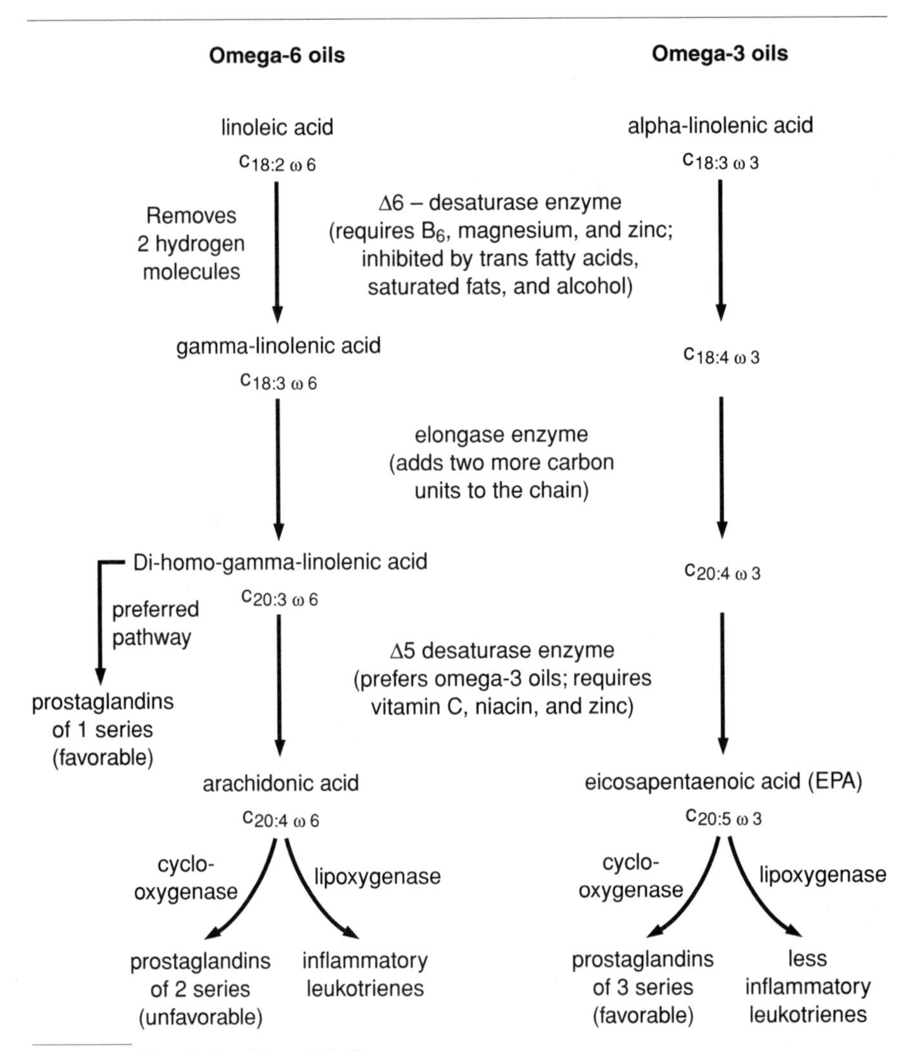

FIGURE 3.1 *Prostaglandin metabolism*

The omega-3 prostaglandin pathway can begin with alpha-linolenic acid, which eventually can be converted to eicosapentaenoic acid, or EPA, which is the precursor to the 3 series prostaglandins. EPA is found preformed in cold-water fish such as salmon, mackerel, and herring. Vegetable oils such as flaxseed and canola, which provide alpha-linolenic acid, can increase body EPA and 3-series prostaglandin levels.

Prostaglandins of the 1 and 3 series are generally viewed as "good" prostaglandins while prostaglandins of the 2 series are viewed as "bad." This labeling is most evident when we look at their effects on platelets. Prostaglandins of the 2 series promote platelet stickiness, which leads to hardening of the arteries, heart disease, and strokes. In contrast, the 1 and 3 series prostaglandins prevent platelets from sticking together, improve blood flow, and reduce inflammation.

Manipulating Prostaglandin Metabolism. By altering the type of dietary oils consumed and stored in cell membranes, prostaglandin metabolism can be manipulated. This manipulation can be extremely powerful in the treatment of inflammation, allergies, high blood pressure, and many other health conditions. The basic goal in most situations is to (1) reduce the level of arachidonic acid inflammation and (2) increase the level of DHGLA and EPA. This goal can best be achieved in most circumstances by reducing the intake of animal foods and supplementing the diet with flaxseed oil.

Although several studies have shown flaxseed oil is not as effective in increasing tissue concentrations of EPA and lowering tissue concentrations of arachidonic acid as fish oils (Cunnane 1991; Nettleton 1991), these studies failed to address an important factor: The studies were all performed on subjects who consumed a diet rich in omega-6 fatty acids. Although the desaturation and elongation enzymes prefer alpha-linolenic acid to the omega-6 oils, in these studies only relatively small amounts of alpha-linolenic acid were converted to EPA because of the much higher concentrations of omega-6 oils. A more recent study was undertaken to determine the potential for dietary flaxseed oil to increase tissue EPA concentrations in healthy human subjects (Mantzioris et al. 1994).

Unlike the previous studies, this study incorporated a diet low in omega-6 oils by restricting the use of other vegetable oils while supplementing the diet daily with 13 grams (approximately 1.5 tablespoons) of flaxseed oil. The results of the study indicated that flaxseed oil supplementation, along with restriction of linoleic acid, will raise tissue EPA levels comparable to fish oil supplementation.

Given the fact that encapsulated fish oils have been shown to contain very high levels of lipid peroxides and are expensive to use at therapeutic dosages (1.8 grams EPA per day), flaxseed oil will more than likely emerge as the preferred source of omega-3 fatty acids in the treatment of atherosclerosis, high blood pressure, and such inflammatory conditions as psoriasis, rheumatoid arthritis, eczema, multiple sclerosis, and ulcerative colitis.

GLA Supplements

Evening primrose, black currant, and borage oil contain gamma-linolenic acid, an omega-6 fatty acid that eventually acts as a precursor to the favorable prostaglandins of the 1 series. Although quite popular, the research on GLA supplements is controversial and is not as strong as the research on omega-3 oils. Studies have shown that over time GLA supplementation will increase tissue arachidonic acid levels while decreasing tissue levels of EPA (Janti 1989). Obviously, this effect is contrary to the treatment goal of trying to reduce inflammation by reducing tissue levels of arachidonic acid and raising levels of EPA.

Another controversial aspect is the fact that because GLA can be formed from linoleic acid, it is difficult to determine to what extent the effects are due to GLA versus linoleic acid. Most sources of GLA are much richer in linoleic acid than GLA. For example, evening primrose contains only 9 percent GLA but 72 percent linoleic acid.

In most cases, oils like safflower and soy that contain high levels of linoleic acid may provide as much benefit as GLA products at a fraction of the cost. The only exceptions to this may be in people with diabetes and those who cannot form GLA from linoleic acid. GLA supplementation in diabetics has been shown to improve nerve function and prevent diabetic nerve disease (Gamma-Linolenic Acid Multicenter Trial Group 1993). However, the dosage required is relatively small (240 to 480 mg of GLA per day).

Once again it appears most people would be better off supplementing their diet with flaxseed oil. Although most of the research on omega-3 oils have featured fish oils rich in eicosapentaenoic acid, EPA can be manufactured in the body from alpha-linolenic acid. Flaxseed oil contains more than twice the amount of omega-3 oil compared to fish oil and is also a good source for linoleic acid. In addition, flaxseed oil may offer other benefits over fish oil and GLA products. Flaxseed oil is also much less expensive than these products.

Achieving Balanced Prostaglandin Synthesis. The production of series 1, 2, and 3 families of prostaglandins is dependent on the type and quality of fats and oils we consume. The challenge lies in consuming an approximate ratio of omega-6 to omega-3 fatty acids that will produce a favorable production of "friendly" series 1 and 3 prostaglandins, while managing the production of the potentially volatile prostaglandin 2 series.

Thanks to scientific review, we now know that the optimal ratio of omega-6 to omega-3 fatty acids is 4:1 (Schlomo and Carasso 1993). This means that to achieve optimal prostaglandin synthesis we should consume four parts omega-6 to one part omega-3. At first this would appear to be good advice. However, most Americans consume ten to twenty times the amount of omega-6 fatty acids that are needed. (Omega-6 fatty acids are the primary fatty acids found in low-quality grocery store oil products and are the primary oil ingredient added to most processed food stuffs). Because of the up to 20:1 omega-6 to omega-3 ratio that we are currently consuming, the last thing we would want would be to increase the amount of omega-6 fatty acids in an attempt to achieve an optimal omega-6 to 3 profile. Instead, a more appropriate approach would be to eliminate the hidden sources of omega-6 oils and supplement the diet with flaxseed oil. The omega-6 to omega-3 ratio for flax is 3:1; this means there are 3 times the amount of omega-3 fatty acids compared to the amount of omega-6 fatty acids. The elimination of low-quality omega-6 fatty acids in your diet and the addition of high-quality flaxseed oil will help you gain the optimal ratio of omega-6 to omega-3 fatty acid profile for balanced and efficient production of prostaglandins and their health-enhancing qualities.

The therapeutic uses of omega-3 oils are extensive due to their ability to correct the underlying imbalance of essential fatty acids in cellular membranes and prostaglandin synthesis. However, most of the research on the therapeutic uses of omega-3 oils has focused on cardiovascular disease, allergic diseases, rheumatoid arthritis, multiple sclerosis, and cancer. These will be the focus of the next four chapters.

References

Cunnane, S. C., et al. (1991). "Alpha-linolenic acid in humans: Direct functional role or dietary precursor." *Nutrition* 7:437–439.

Gamma-Linolenic Acid Multicenter Trial Group. (1993). "Treatment of diabetic neuropathy with gamma-linolenic acid." *Diabetes Care* 16:8–15.

Janti, J. (1989). "Evening primrose oil in rheumatoid arthritis: Changes in serum lipids and fatty acids." *Annals Rheum Dis* 48:124–127.

Mantzioris, E., et al. (1994). "Dietary substitution with alpha-linolenic acid-rich vegetable oil increases eicosapentaenoic acid concentrations in tissues." *Am J Clin Nutr* 59:1304–1309.

Nettleton, J. A. (1991). "Omega-3 fatty acids: Comparison of plant and seafood sources in human nutrition." *J Am Diet Assoc* 91:331–337.

Schlomo, Y, and Carasso, R. L. (1993). "Modulation of learning, pain thresholds, and thermoregulation in the rat by preparations of free purified alpha-linolenic and linoleic acids: Determination of the optimal w3-to-w6 ratio." *Proc Natl Acad Sci* 90:10345–10347.

OMEGA-3 OILS IN LOWERING CHOLESTEROL AND BLOOD PRESSURE

4

Most cardiovascular diseases are the result of atherosclerosis, or hardening of the artery walls, due to a buildup of plaque containing cholesterol, fatty material, and cellular debris. For example, if the blood flow through the coronary arteries, which supply the heart with oxygen and nutrients, is severely blocked because of the buildup of cholesterol-containing plaque, severe damage or death to the heart muscle may occur. Heart attacks, strokes, and other cardiovascular diseases related to atherosclerosis are responsible for roughly 43 percent of all deaths in the United States.

Atherosclerosis is a degenerative condition of the arteries characterized by accumulation of lipids (mainly cholesterol, usually complexed to proteins, and cholesterol esters) within the artery. The atherosclerotic plaque, or atheroma, represents the endpoint of a complex, insidious process. Although any artery may be affected, the aorta, coronary arteries, and arteries supplying the brain are most frequently involved.

The Role of Cholesterol in Atherosclerosis

The first step in the prevention and treatment of heart disease and strokes is the reduction of blood cholesterol levels. The evidence overwhelmingly demonstrates that elevated cholesterol levels are deadly. However, not all cholesterol is bad; it serves many vital functions in the body, including the manufacture of sex hormones and bile acids. Without cholesterol many body processes would not function properly. Cholesterol is transported in the blood by molecules known as *lipoproteins*. Cholesterol bound to low-density lipoprotein, or LDL, is often referred to as the "bad" cholesterol while cholesterol bound to high-density lipoprotein, or HDL, is the "good" cholesterol. LDL cholesterol increases the risk for heart disease, strokes, and high blood pressure; HDL cholesterol actually protects against heart disease.

LDL transports cholesterol to the tissues. HDL, on the other hand, transports cholesterol to the liver for metabolism and excretion from the body. The LDL to HDL ratio (also referred to as the cardiac risk factor ratio) largely determines whether cholesterol is being deposited into tissues or broken down and excreted. The risk for heart disease can be reduced dramatically by lowering LDL cholesterol while simultaneously raising HDL cholesterol levels. Research has shown that for every 1 percent drop in the LDL cholesterol level, the risk for a heart attack drops by 2 percent. Conversely, for every 1 percent increase in HDL levels the risk for a heart attack drops 3 to 4 percent.

In addition to maintaining the proper cholesterol level, it is also important to keep the level of triglycerides in the proper range. Following are the recommended levels of blood cholesterol and triglycerides:

Total cholesterol . . . . less than 200 mg/dl
LDL cholesterol . . . . less than 130 mg/dl
HDL cholesterol . . . . greater than 35 mg/dl
LDL to HDL ratio . . . less than 4:5
Triglycerides . . . . 50 to 150 mg/dl

Omega-3 Oils and Cholesterol Levels

Omega-3 oils have been shown in hundreds of studies to lower cholesterol and triglyceride levels (Bierenbaum et al. 1993; Bjerve et al. 1992; Schmidt and Dyerberg 1994; Simopoulos 1991; Von Schacky 1987). The majority of these studies have featured EPA- and DHA-rich fish oils, but flaxseed oil can produce similar benefits because it contains linolenic acid, an omega-3 oil that the body can convert to eicosapentaenoic acid (EPA) (see page 23). Linolenic acid exerts many of the same effects as EPA.

Though there is much evidence documenting the beneficial effects of increasing the intake of fish oils in lowering blood cholesterol levels, the question remains whether fish oils should be taken as a supplement or whether the intake of fish should be increased in the diet. In an effort to resolve this question, a recent five-week study on twenty-five men with high cholesterol levels compared the effects of eating an equivalent amount of fish oil from whole fish versus a fish oil supplement (Cobias et al. 1991). Although total cholesterol levels were unchanged in both groups, both fish and fish oil supplements lowered triglycerides and raised HDL cholesterol. However, dietary fish produced a much greater effect than the fish oil supplement on reducing platelet stickiness and preventing clot formation.

Once platelets adhere to each other, or aggregate, they release potent compounds that dramatically promote the formation of the atherosclerotic plaque, or a clot forms that can get stuck in small arteries and produce a heart attack or stroke. While saturated fats and cholesterol increase platelet aggregation, omega-3 oils have the opposite effect. As described in Chapter 3, these effects are mediated through improved prostaglandin metabolism.

While both fish consumption and fish oil supplementation produce desirable effects on cholesterol, fish consumption is more effective in preventing cardiovascular disease.

Because commercially available fish oils contain very high levels of lipid peroxides and greatly stress antioxidant defense mechanisms, it makes more sense to rely on cold-water fish and flaxseed oil for the omega-3 oils rather than on fish oil capsules (Fritshe and Johnston 1988; Harats et al. 1991; Shukla and Perkins 1991).

Flaxseed oil will provide a significant cost savings as well. The dosages found to be effective in lowering cholesterol levels when using fish oil supplements range from 5 to 15 grams per day. Since most commercial products contain 500 mg of fish oil per capsule, this means a daily dose of 10 to 30 capsules and an average monthly cost of $40 to $100. At a similar dosage, flaxseed oil would cost between $6 and $18 per month.

Diets High in Omega-3 Oils Prevent Heart Attacks

Population studies have demonstrated that people who consume a diet rich in omega-3 oils from either fish or vegetable sources have a significantly reduced risk of developing heart disease (Kromann and Green 1980; Kromhout, Bosscheiter, and De Lezenne-Coulander 1985). Furthermore, results from autopsy studies have shown that the highest degree of coronary artery disease is found in individuals with the lowest concentration of omega-3 oils in their fat tissues. Conversely, individuals with the lowest degree of coronary artery disease had the highest concentration of omega-3 oils (Seidelin, Myrup, and Fischer-Hansen 1992).

What these studies show is that people consuming a diet rich in omega-3 oils can prevent heart attacks. But what about people who already have significant heart disease? Can a diet rich in omega-3 oils prevent future heart attacks? The answer is yes!

People who have experienced a heart attack and live through it are extremely likely to experience another. Several studies have sought to determine whether dietary modifications can prevent recurrence. As of 1995, only three studies have shown that dietary modifications are effective.

Diet and lifestyle are not only protective against heart disease, they can dramatically reverse the blockage of clogged arteries. The most famous of the three studies showing this effect is the Lifestyle Heart Trial conducted by Dr. Dean Ornish (1990). Subjects with heart disease were divided into a control group and an experimental group. The control group received regular medical care while the experimental group was asked to eat a low-fat vegetarian diet for at least one year. The diet included fruits, vegetables, grains, legumes, and soybean products. Subjects were allowed to consume as many calories as they wished. No animal products were allowed except egg white and one cup per day of non-fat milk or yogurt. The diet contained approximately 10 percent fat, 15 to 20 percent protein, and 70 to 75 percent carbohydrates (predominantly complex carbohydrates from whole grains, legumes, and vegetables).

The experimental group was also asked to perform such stress reduction techniques as breathing exercises, stretching exercises, meditation, imagery, and other relaxation techniques for an hour each day and to exercise at least three hours a week. At the end of the year, the subjects in the experimental group showed significant overall regression of atherosclerosis of the coronary blood vessels. In contrast, subjects in the control group, who were being treated with regular medical care and following the standard American Heart Association diet, showed progression of their disease—they got worse. Ornish (1990) states, "This finding suggests that conventional recommendations for patients with coronary heart disease (such as a 30 percent fat diet) are not sufficient to bring about regression in many patients."

Strict vegetarianism may not be as important as consuming a diet high in fiber and complex carbohydrates and low in saturated fat and cholesterol, but it is well established that a vegetarian diet has been shown to be quite effective in lowering cholesterol levels and blood pressure, reducing the risk for atherosclerosis. Such a diet is rich in protective factors such as fiber, essential fatty acids (including higher levels of alpha-linolenic acid), vitamins, and minerals (including potassium and magnesium).

Two other studies showing that diet can prevent further heart attacks in patients suffering a first heart attack highlight the importance of omega-3 fatty acids and the ineffectiveness of the standard American Heart Association's dietary recommendations. In the Dietary and Reinfarction Trial (DART) when the intake of omega-3 fatty acids (from fish) was increased, future heart attacks were reduced (Burr et al. 1989). The Lyon Diet Heart Study determined that increasing the intake of alpha-linolenic acid offers the same degree of protection as increased fish intake (de Lorgeril et al. 1994). The diet used in the Lyon Heart Study is often referred to as the "Cretan Diet."

The Japanese who inhabit Kohama Island and the inhabitants of Crete, the two populations with the lowest rate of heart attacks, have a relatively high intake of alpha-linolenic acid (Kagawa et al. 1982; Sandker et al. 1993). Cretans typically have a three-fold higher concentration of alpha-linolenic acid compared to members of other European countries due to their frequent consumption of walnuts and purslane. Another important dietary factor in both the Kohamans and Cretans is their use of oils containing oleic acid—canola and olive, respectively. LDL cholesterol largely composed of oleic acid is less susceptible to peroxidation. Although the oleic content of the diet offers some degree of protection, the rate of heart attacks in the Kohamans and Cretans is much lower than populations that consume only oleic acid sources and little alpha-linolenic acid. The intake of alpha-linolenic acid is viewed as a more significant protective factor.

Effects of Omega-3 Oils on Fibrinogen. Elevated fibrinogen levels are a major risk factor for cardiovascular disease. Fibrinogen, a protein involved in the clotting system, plays an important role in promoting atherosclerosis, such as acting as a co-factor for platelet aggregation, determining the viscosity of blood, and stimulating the formation of the atherosclerotic plaque.

Early clinical studies stimulated detailed epidemiological investigations on the possible link between fibrinogen and cardiovascular disease (Ernst, 1994). The first study was the Northwick Park Heart Study in England. This study involved 1,510 men ages forty to sixty-four who were randomly recruited and tested for a range of clotting factors, including fibrinogen. At the four-year follow-up there was a stronger association between cardiovascular deaths and fibrinogen levels than that for cholesterol. This association has been confirmed in five other prospective epidemiological studies.

The clinical significance of these findings can be summarized as follows:

1. Fibrinogen levels should be determined and monitored in patients with or at high risk for coronary heart disease or stroke.
2. Natural therapies (e.g., omega-3 oils) designed to promote fibrinolysis may offer significant benefit in the prevention of heart attacks and strokes.

Though omega-3 oils lower fibrinogen levels, omega-6 oils do not (Radack, Deck, and Huster 1990). This difference may explain some of the results noted in population and clinical studies such as the DART and Lyon Diet Heart Study.

Omega-3 Oils and Blood Pressure. Each time the heart beats it sends blood coursing through the arteries. The peak reading of the pressure exerted by this contraction is the systolic pressure. Between beats the heart relaxes and blood pressure drops. The very lowest reading is referred to as the diastolic pressure. A normal blood pressure reading for an adult is 120 (systolic) / 80 (diastolic).

High blood pressure, or hypertension, refers to a reading of greater than 140/90. An elevated blood pressure is one of the major risk factors for a heart attack or stroke. Because heart disease and strokes account for over 43 percent of all deaths in the United States, it is very important to keep the blood pressure in the normal range. Over 60 million Americans have high blood pressure. Again, dietary factors appear to be the primary reason.

Besides attaining ideal body weight, perhaps the most important dietary recommendation is to increase the consumption of plant foods in the diet. A primarily vegetarian diet typically contains more potassium, complex carbohydrates, essential fatty acids, fiber, calcium, magnesium, and vitamin C and less saturated fat and refined carbohydrates, all of which have a favorable influence on blood pressure.

Increasing the intake of omega-3 fatty acids can also lower blood pressure. Over sixty double-blind studies have demonstrated that either fish oil supplements or flaxseed oil are very effective in lowering blood pressure (Appel et al. 1993; Schmidt and Dyerberg 1994; Singer 1992). Although the fish oils have typically produced a more pronounced effect than flaxseed oil, because the dosage of fish oils used in the studies was quite high (e.g., equal to 100 fish oil capsules a day), flaxseed oil is the better choice for lowering blood pressure, especially when cost effectiveness is considered.

Along with reducing the intake of saturated fat, one tablespoon per day of flaxseed oil should drop both the systolic and diastolic readings by up to 9 mm Hg (Chan, Bruce, and McDonald 1991). One study found that for every absolute 1 percent increase in body alpha-linolenic acid content there was a decrease of 5 mm Hg in the systolic, diastolic, and mean blood pressure (Berry and Hirsch 1986).

Summing Up

The beneficial effects of omega-3 oils in protecting as well as treating cardiovascular disease are quite obvious. Omega-3 oils impact numerous factors linked to heart attacks and strokes: They lower LDL-cholesterol levels and triglycerides, inhibit excessive platelet aggregation, lower fibrinogen levels, and lower both systolic and diastolic blood pressure in individuals with high blood pressure. Flaxseed oil offers the most cost-effective and beneficial method for increasing the intake of omega-3 oils in the diet.

References

Appel, L. J., et al. (1993). "Does supplementation of diet with 'fish oil' reduce blood pressure? A meta-analysis of controlled clinical trials." *Arch Intern Med* 153:1429–1438.

Berry, E. M., and Hirsch, J. (1986). "Does dietary linolenic acid influence blood pressure?" *Am J Clin Nutr* 44:336–340.

Bierenbaum, M. L., et al. (1993). "Reducing atherogenic risk in hyperlipemic humans with flax seed supplementation: A prelimary report." *J Am Coll Nutr* 12:501–504.

Bjerve, K. S., et al. (1992). "Clinical studies with alpha-linolenic acid and long chain n-3 fatty acids." *Nutrition* 8:130–132.

Burr, M. L., et al. (1989). "Effects of changes in fat, fish, and fiber intakes on death and myocardial reinfarction: Diet and reinfarction trial (DART)." *Lancet* 334:757–761.

Chan, J. K., Bruce, V. M., and McDonald, B. E. (1991). "Dietary-alpha-linolenic acid is as effective as oleic acid and linoleic acid in lowering blood cholesterol in normolipidemic men." *Am J Clin Nutr* 53:1230–1234.

Cobias, L., et al. (1991). "Lipid, lipoprotein, and hemostatic effects of fish versus fishoil w-3 fatty acids in mildly hyperlipidemic males." *Am J Clin Nutr* 53:1210–1216.

de Lorgeril, M., et al. (1994). "Mediterranean alpha-linolenic acid-rich diet in secondary prevention of coronary heart disease." *Lancet* 343:1454–1459.

Ernst, E. (1994). "Fibrinogen: An important risk factor for atherothrombotic diseases." *Annals Med* 26:15–22.

Fritshe, K. L., and Johnston, P. V. (1988). "Rapid autoxidation of fish oil in diets without added antioxidants." *J Nutr* 118:425–426.

Harats, D., et al. (1991). "Fish oil ingestion in smokers and nonsmokers enhances peroxidation of plasma lipoproteins." *Atherosclerosis* 90:127–139.

Kagawa, Y., et al. (1982). "Eicosapolyenoic acids of serum lipids of Japanese Islanders with low incidence of cardiovascular diseases." *J Nutr Sci Vitaminol* 28:441–453.

Kromann, N., and Green, A. (1980). "Epidemiological studies in the Upernavik district, Greenland." *Acta Med Scand* 208:401–406.

Kromhout, D., Bosscheiter, E. B., and De Lezenne-Coulander, C. (1985). "Inverse relation between fish oil consumption and 20-year mortality from coronary heart disease." *New Engl J Med* 312:1205–1209.

Ornish, D., et al. (1990). "Can lifestyle changes reverse coronary heart disease?" *Lancet* 336:129–133.

Radack, K., Deck, C., and Huster, G. (1990). "The comparative effects of n-3 and n-6 polyunsaturated fatty acids on plasma fibrinogen levels: A controlled clinical trial in hypertriglyeridemic subjects." *J Am Coll Nutr* 9:352–357.

Sandker, G. N., et al. (1993). "Serum cholesterol ester fatty acids and their relation with serum lipids in elderly men in Crete and the Netherlands." *Eur J Clin Nutr* 47:201–208.

Schmidt, E. B., and Dyerberg, J. (1994). "Omega-3 fatty acids. Current status in cardiovascular medicine." *Drugs* 47:405–424.

Seidelin, K. N., Myrup, B., and Fischer-Hansen, B. (1992). "n-3 fatty acids in adipose tissue and coronary artery disease are inversely correlated." *Am J Clin Nutr* 55:1117–1119.

Shukla, V. K. S., and Perkins, E. G. (1991). "The presence of oxidative polymeric materials in encapsulated fish oils." *Lipids* 26:23–26.

Simopoulos, A. P. (1991). "Omega-3 fatty acids in health and disease and in growth and development." *Am J Clin Nutr* 54:438–463.

Singer, P. (1992). "Alpha-linolenic acid vs. long-chain fatty acids in hypertension and hyperlipidemia." *Nutrition* 8:133–135.

Von Schacky, C. (1987). "Prophylaxis of atherosclerosis with marine omega-3 fatty acids: A comprehensive strategy." *Annals Int Med* 107:890–899.

OMEGA-3 OILS IN ALLERGIC AND INFLAMMATORY CONDITIONS

5

Fatty acids are important mediators of allergy and inflammation through their ability to form inflammatory prostaglandins, thromboxanes, and leukotrienes. Altering dietary oil intake can significantly increase or decrease inflammation, depending on the type of oil being increased. The overall goal is to decrease tissue levels of arachidonic acid while simultaneously increasing the tissue levels of omega-3 fatty acids.

Vegetarian diets are often beneficial in the treatment of many chronic allergic and inflammatory conditions, presumably as a result of decreasing the availability of arachidonic acid for conversion to inflammatory prostaglandins and leukotrienes while simultaneously supplying linoleic and linolenic acids. These fatty acids lead to the formation of prostaglandins which actually inhibit inflammation.

Essential Fatty Acids in Allergic and Inflammatory Conditions

Numerous clinical studies have demonstrated a therapeutic effect when supplementing the diet with essential fatty acids in the treatment of many chronic allergic and inflammatory diseases, including rheumatoid arthritis, asthma, eczema, psoriasis, lupus, and ulcerative colitis. Particularly beneficial are the omega-3 fatty acids.

To illustrate the power of dietary changes in fatty acids to improve a severe chronic degenerative disease, let's examine the clinical studies of essential fatty acids in the treatment of rheumatoid arthritis.

GLA Supplements in Rheumatoid Arthritis

GLA supplements (i.e., black currant, borage, and evening primrose oil) are often recommended by nutritionally oriented physicians in the treatment of such inflammatory conditions as rheumatoid arthritis and eczema. However, these oils may actually produce a negative effect on tissue fatty acid patterns by raising the level of arachidonic acid and reducing the level of omega-3 fatty acids (Jantti et al. 1989).

Certain studies have shown some benefit with GLA supplementation in the treatment of rheumatoid arthritis (Belch et al. 1988; Brzeski, Madhok, and Capell 1991). The key factor appears to be whether or not subjects are allowed to take their anti-inflammatory drugs. These drugs inhibit the formation of inflammatory prostaglandins and mask the negative effects of the altered tissue fatty acid profile produced by GLA supplements.

Although positive results have been reported with GLA supplementation, closer examination is required. For example, in one double-blind study, thirty-seven patients with rheumatoid arthritis were given either GLA (1.4 grams daily) or a placebo for twenty-four weeks. GLA supplementation reduced the number of tender joints by 36 percent, the tender joint score by 45 percent, swollen joint count by 28 percent, and the swollen joint score by 41 percent (Levanthal et al. 1993). In contrast, no patients in the placebo group showed significant improvement in any measure. The superficial results of this study indicate that borage oil may be useful in reducing the inflammatory process of rheumatoid arthritis. However, in this study the subjects continued to take their anti-inflammatory drugs and probably masked the detrimental effects on tissue arachidonic acid and EPA levels.

The recommended daily dosage for GLA in the treatment of rheumatoid arthritis is 1.4 grams. Since evening primrose oil is 9 percent GLA, this means that approximately thirty-one 500 mg capsules of evening primrose oil would have to be consumed each day. This dosage would typically cost a person roughly $100 per month. Taking less than the recommended dosage is not likely to produce benefit. For several reasons, including cost, flaxseed oil appears to be a better choice.

Fish Oil Supplements in Rheumatoid Arthritis

The studies of fish oil supplementation in the treatment of rheumatoid arthritis have demonstrated far better and more consistent results than the studies with GLA supplementation. The first double-blind, placebo-controlled study of rheumatoid arthritis patients using 1.8 grams of EPA a day showed less morning stiffness and tender joints (Kremer et al. 1985). These results led to considerable scientific interest as well as numerous popular press accounts of the possible benefits of fish oil for allergic and inflammatory conditions.

Over a dozen follow-up studies have consistently demonstrated positive benefits (Cleland et al. 1988; Kremer et al. 1987, 1990; Lau et al. 1991; Magaro et al. 1988; Nielsen et al. 1992; Sperling et al. 1987; van der Temple et al. 1990). As well as improvements in symptoms (morning stiffness and joint tenderness), fish oil supplementation has produced favorable changes in suppressing the production of inflammatory compounds secreted by white blood cells.

Although the results of these studies are impressive, all the studies were less than one year. In order to properly assess the beneficial effect of any treatment of rheumatoid arthritis, it is important to evaluate patients over a longer period of time, as the condition is associated with ups and downs in symptom severity. Recently, a one-year study of fish oil supplementation in the treatment of rheumatoid arthritis was completed. The results clearly indicated that supplementation with 2.6 grams per day of omega-3 oil (six 1-gram capsules of fish oil per day) resulted in significant clinical benefit and led to significant reductions in the need for drug therapy. The results of this long-term study provide further validation of the short-term studies.

Although there are no studies of flaxseed oil in the treatment of rheumatoid arthritis, there is considerable evidence that suggests that flaxseed oil can produce the same kind of results as fish oils. Supplementation with alpha-linolenic acid can produce the same kind of changes as fish oil (EPA) supplementation (Mantzioris et al. 1994). Furthermore, several human and animal studies have demonstrated that flaxseed oil supplementation can inhibit the autoimmune reaction as effectively as EPA (Kelley 1992).

Again, look at the cost. In order to provide benefit in the treatment of rheumatoid arthritis, the dosage of fish oil must provide 1.8 grams of EPA each day. This dosage equates to approximately ten 1-gram capsules daily with an approximate cost of $70 to $90 a month. On the other hand, it takes 1.5 tablespoons of flaxseed oil to produce similar effects on tissue arachidonic acid and EPA levels and costs $12 to $18 a month.

Food Allergies

Food allergies play a major role in many chronic allergic and inflammatory conditions (Brostoff and Challacombe 1987). A food allergy occurs when there is an adverse reaction to the ingestion of a specific food. The actual symptoms produced during an allergic response depend on the location of the immune system activation, the mediators of inflammation involved, and the sensitivity of the tissues to these mediators.

Although there are laboratory tests available that identify food allergies, many physicians believe that oral food challenge is the best way of diagnosing adverse reactions to foods. The most popular method involves using an elimination diet. The patient is placed on a limited diet; commonly eaten foods are eliminated and replaced with either

SYMPTOMS AND DISEASES COMMONLY ASSOCIATED WITH FOOD	
System	**Symptoms and Diseases**
Gastrointestinal	Canker sores, celiac disease, chronic diarrhea, stomach ulcer, gas, gastritis, irritable colon, malabsorption, ulcerative colitis
Genitourinary	Bedwetting, chronic bladder infections, kidney disease
Immune	Chronic infections, frequent ear infections
Brain	Anxiety, depression, hyperactivity, inability to concentrate, insomnia, irritability, mental confusion, personality change, seizures
Musculoskeletal	Bursitis, joint pain, low back pain
Respiratory	Asthma, chronic bronchitis, wheezing
Skin	Acne, eczema, hives, itching, skin rash
Miscellaneous	Irregular heart beats, edema, fainting, fatigue, headache, hypoglycemia, itchy nose or throat, migraines, sinusitis

hypoallergenic foods rarely eaten or special hypoallergenic meal replacement formulas. The standard elimination diet (also known as an oligoiantigenic diet) consists of lamb, chicken, potato, rice, banana, apple, and a vegetable in the cabbage family (e.g., cabbage, Brussels sprouts, broccoli).

After the elimination diet or fast, the patient's individual foods are reintroduced. A "new" food item is reintroduced to the diet every second day. Reintroduction of sensitive foods will typically produce a more severe or recognizable symptom than before. A detailed record must be maintained describing when foods were reintroduced and what symptoms appeared upon reintroduction. If there is an increase in pain, stiffness, or joint swelling within two to forty-eight hours of a reintroduction of a food item, this food must be omitted from the diet for at least seven days before being reintroduced a second time. If the food causes worsening of symptoms after the second reintroduction, it should probably be omitted permanently from the diet.

To highlight just how effective elimination of food allergies can be in a chronic inflammatory condition, let's look at a thirteen-month study of two groups of patients suffering from rheumatoid arthritis conducted in Norway at the Oslo Rheumatism Hospital (Kjeldsen-Kragh et al. 1991). The treatment group followed a therapeutic diet; the control group was allowed to eat as they wished. Both groups started the study by visiting a health farm, or spa, for four weeks.

The members of the treatment group began their therapeutic diet by fasting for seven to ten days and then began following a special diet. Dietary intake during the fast consisted of herbal teas, garlic, vegetable broth, decoction of potatoes and parsley, and carrot, beet, and celery juices. After the fast the patients reintroduced a new food item every second day. If they noticed an increase in pain, stiffness, or joint swelling within two to forty-eight hours, this item was omitted from the diet for at least seven days before being reintroduced a second time. If the food caused worsening of symptoms after the second reintroduction, it was omitted permanently from the diet.

The results of the study show that short-term fasting followed by a vegetarian diet results in "a substantial reduction in disease activity" in many patients. The results also indicated a therapeutic benefit beyond elimination of food allergies alone. Kjeldsen-Kragh and his colleagues suggest that the additional improvements are due to changes in dietary fatty acids.

Summing Up

Dietary fatty acids play a major role in the allergic and inflammatory response because they are converted in the body to prostaglandins and leukotrienes—important mediators of inflammation. By restricting the intake of omega-6 fatty acids and supplementing the diet with flaxseed oil, a favorable effect on the tissue levels of arachidonic acid and EPA occurs. As a result, allergic and inflammatory processes are reduced and the condition improves.

References

Belch, J. F., et al. (1988). "Effects of altering dietary essential fatty acids on requirements for non-steroidal anti-inflammatory drugs in patients with rheumatoid arthritis: A double-blind placebo controlled study." *Annals Rheum Dis* 47:96–104.

Brostoff, J., and Challacombe, S. J. (eds.) (1987). *Food Allergy and Intolerance.* W. B. Saunders, Philadelphia, PA.

Brzeski, M., Madhok, R., and Capell, H. A. (1991). "Evening primrose oil in patients with rheumatoid arthritis and side effects of non-steroidal anti-inflammatory drugs." *Br J Rheumatol* 30:371–372.

Cleland, L. G., et al. (1988). "Clinical and biochemical effects of dietary fish oil supplements in rheumatoid arthritis." *J Rheumatol* 15:1471–1475.

Jantti, J., et al. (1989). "Evening primrose oil in rheumatoid arthritis: Changes in serum lipids and fatty acids." *Annals Rheum Dis* 48:124–127.

Kelley, D. S. (1992). "Alpha-linolenic acid and immune response." *Nutrition* 8:215–217.

Kjeldsen-Kragh, J., et al. (1991). "Controlled trial of fasting and one-year vegetarian diet in rheumatoid arhtritis." *Lancet* 338:899–902.

Kremer, J., et al. (1985). "Effects of manipulation of dietary fatty acids on clinical manifestation of rheumatoid arthritis." *Lancet* (January 26):184–187.

Kremer, J., et al. (1987). "Fish-oil supplementation in active rheumatoid arthritis: A double-blinded, controlled cross-over study." *Annals Intern Med* 106:497–502.

Kremer, J. M., et al. (1990). "Dietary fish oil and olive oil supplementation in patients with rheumatoid arthritis." *Arthritis Rheum* 33:810–20.

Lau, C. S., et al. (1991). "Maxepa on nonsteroidal anti-inflammatory drug usage in patients with mild rheumatoid arthritis." *Br J Rheumatol* 30:137.

Levanthal, L. J., et al. (1993). "Treatment of rheumatoid arthritis with gammalinoleic acid." *Annals Int Med* 119:867–873.

Magaro, M., et al. (1988). "Influence of diet with different lipid composition on neutrophil composition on neutrophil chemiluminescence and disease activity in patients with rheumatoid arthritis." *Annals Rheum Dis* 47:793–796.

Mantzioris, E., et al. (1994). "Dietary substitution with alpha-linolenic acid-rich vegetable oil increases eicosapentaenoic acid concentrations in tissues." *Am J Clin Nutr* 59:1304–1309.

Nielsen, G. L., et al. (1992). "The effects of dietary supplementation with n-3 polyunsaturated fatty acids in patients with rheumatoid arthritis: A randomized, double-blind trial." *Eur J Clin Invest* 22:687–691.

Sperling, R., et al. (1987). "Effects of dietary supplementation with marine fish oil on leukocyte lipid mediator generation and function in rheumatoid arthritis." *Arthritis Rheum* 30:988–997.

van der Temple, H., et al. (1990). "Effects of fish oil supplementation in rheumatoid arthritis." *Annals Rheum Dis* 49:76–80.

OMEGA-3 OILS IN MULTIPLE SCLEROSIS

6

Multiple sclerosis (MS) is a debilitating condition of progressive nervous system disturbance. The early symptoms of multiple sclerosis may include

- muscular symptoms—feeling of heaviness, weakness, leg dragging, stiffness, tendency to drop things, clumsiness
- sensory symptoms—tingling, pins-and-needles sensation, numbness, dead feeling, bandlike tightness, electrical sensations
- visual symptoms—blurring, fogginess, haziness, eyeball pain, blindness, double vision
- vestibular symptoms—light-headedness, feeling of spinning, sensation of drunkenness, nausea, vomiting
- genitourinary symptoms—incontinence, loss of bladder sensation, loss of sexual function

Despite considerable research, there are still many questions about MS. Mainstream medicine has become almost obsessed with finding a viral cause for this disease, although most current work suggests immune disturbances. In MS, the myelin sheath that surrounds nerves is destroyed. For this reason MS is classified as a "demyelinating" disease. Zones of demyelination (plaques) vary in size and location within the spinal cord. Symptoms correspond in a general way to the distribution of the plaques.

In about two-thirds of the cases, onset is between ages twenty and forty (rarely is the onset after fifty), and women are affected slightly more often than males (60 percent female; 40 percent male). Though the cause of MS is still undetermined, many causative factors have been proposed, including viruses, autoimmune factors, and diet.

The Role of Essential Fatty Acids in MS

People with MS are thought to have a defect in essential fatty absorption or transport, which results in a functional deficiency state. In addition, because consumption of saturated fats increases the requirements of essential fatty acids, a relative deficiency state exists in many cases even if the level of essential fatty acids in the diet is adequate in normal circumstances. Central to this defect in essential fatty acid metabolism in MS are deficiencies of the omega-3 oils, which play a critical role in the structure and function of myelin. A deficiency of alpha-linolenic acid and other omega-3 fatty acids can result in permanently impairing the formation of normal myelin (Cumane 1991; Simopoulos 1991).

Roy Swank's MS Diet

Dr. Roy Swank (1977), Professor of Neurology, University of Oregon Medical School, has provided convincing evidence that a diet low in saturated fats and high in essential fatty acids maintained over many years (one study lasted more than thirty-four years) tends to halt the disease process. Swank began successfully treating patients with his low-fat diet in 1948. His diet recommendations are to (1) eliminate butter, hydrogenated oils (margarine and shortening), and intake no more than 15 grams of animal fat per day; (2) consume daily 40 to 50 grams, or approximately 3 to 4 tablespoons, of polyunsaturated vegetable oils; (3) take at least 1 teaspoon of cod-liver oil daily; (4) eat a normal amount of protein, mostly from vegetables, nuts, fish, white meat of turkey and chicken (skin removed), and lean meat; and (5) eat fish three or more times a week.

The results of the thirty-four-year study in MS patients conducted by Swank from 1949 to 1984 are astounding. Minimally disabled patients who followed his dietary recommendations experienced little disease progression, if at all. Only 5 percent of those in the study did not survive, but 80 percent of those who failed to follow the diet recommendations did not survive the study period. The moderately and severely disabled patients who followed the dietary recommendations also did far better than the group that didn't. In addition to dramatically reducing the death rate, the diet was shown to prevent worsening of the disease and greatly reduced fatigue.

Swank's results in MS along with numerous other studies in other chronic degenerative diseases support the critical need to alter the production of prostaglandins and related compounds in these conditions by changing the source of dietary fatty acids. Swank's diet appears to be a good recommendation for virtually any chronic inflammatory or allergic condition. At the very least, such dietary manipulation will lessen the dosage of corticosteroids required.

Currently, it is thought that the beneficial effects of Swank's diet are a result of (1) decreasing the aggregation of blood platelets, (2) decreasing an autoimmune response, (3) a reduction in saturated fat intake, and (4) normalizing the decreased essential fatty acid levels found in the serum, red blood cells, platelets, and, perhaps most important, the cerebrospinal fluid in patients with MS.

Linoleic Acid Supplementation in MS

Linoleic acid has been investigated as a treatment for MS in three double-blind trials (Bates et al. 1978; Millar et al. 1973; Paty et al. 1978). Although the results of the studies are mixed, combined analysis indicates that patients supplemented with linoleic acid have a smaller increase in disability and reduced severity and duration of relapses than do controls. These studies used a sunflower seed oil at dosage of a little more than one tablespoon per day.

The authors believe that better results would have been attained in the double-blind studies if dietary saturated fatty acids had been restricted, larger amounts of essential fatty acids had been used (at least two tablespoons per day), and the studies had been of longer duration (one study found that normalization of fatty acid levels required at least two years of supplementation). Furthermore, the authors believe that even better results would have been attained if flaxseed oil had been used. Flaxseed oil not only contains linoleic acid but also alpha-linolenic acid. Linolenic acid has a greater effect on reducing platelet aggregation than linoleic acid and is required for normal central nervous system composition.

Benefits of Omega-3 Fatty Acids in MS

There appears to be a strong rationale for supplementation with omega-3 fatty acids in the treatment of MS. Although no direct clinical investigation has been done, one of the key recommendations in Swank's diet was the regular consumption of fish and cod-liver oil. Omega-3 oils are essential in the formation of normal collagen. Swank's results, along with this essential role, provides ample reasons to supplement with omega-3 oils in the treatment of MS. There are also some interesting geographical features on the distribution of MS that seem to indicate that omega-3 oils may help prevent or treat MS.

Some of the first investigations into diet and MS centered around trying to explain why inland farming communities in Norway had a higher incidence of MS than areas near the coastline (Swank et al. 1952). It was discovered that the diets of the farmers were much higher in animal and dairy products than the diets of the coastal dwellers, and the coastal dwellers' diet had much higher levels of cold-water fishes rich in omega-3 oils. Because animal and dairy products are much higher in saturated fatty acids and lower in polyunsaturated fatty acids than fish, researchers explored this association in greater detail. Subsequent studies have upheld a strong association between a diet rich in animal and dairy products and the incidence of MS (Swank and Pullen 1977). In other words, when the intake of saturated fats and animal foods is high, so is the rate of MS.

Unfortunately, researchers have been slow in determining the role the level of omega-3 oils in the diet can play in reducing the risk for MS. In addition to the studies in Norway, it is interesting to note that the incidence of MS is quite low in Japan where consumption of marine foods, seeds, and fruit oil rich in omega-3 oils is quite high.

GLA and MS

It has been suggested that gamma-linolenic acid (GLA) may be more effective than linoleic acid alone because of its more ready incorporation into brain lipids and its possibly greater effect on immune function (Horrobin 1979). However, due to its cost and the fact that relatively large amounts of the product would have to be consumed to exert a therapeutic effect, supplementation with GLA is not indicated at this time. One study demonstrated that daily supplementation with only 340 mg of gamma-linoleic and 2.92 grams of linoleic acid (the ratio found in evening primrose oil) had no effect on the clinical course in MS patients (Bates et al. 1978). In the same study, those receiving 23 grams of linoleic acid demonstrated reduced frequency and severity of acute attacks, even though the study was conducted for only twenty-four months.

Additional Recommendations

Many studies have demonstrated reduced antioxidant enzyme activity in MS (Jensen, Gissel-Nielsen, and Clausen 1980; Mazzella et al. 1983; Shukla, Jensen, and Clausen 1977; Szeinberg et al. 1979; Wikstrom, Westermarck, and Palo 1976). The enzyme involved is the selenium-containing enzyme glutathione peroxidase (GSH-Px). The glutathione peroxidase enzyme is intricately involved in the protection of the myelin sheath from free radical damage. Therefore, a decreased activity level would leave the myelinated sheath particularly sensitive to damage.

Because geographical areas low in selenium often overlap high-rate areas for MS, it is natural to speculate that there may be a correlation between selenium levels, glutathione peroxidase activity, and MS. Initial studies seem to support this correlation (Jensen, Gissel-Nielsen, and Clausen 1980; Mazzella et al. 1983; Shukla, Jensen, and Clausen 1977; Szeinberg et al. 1979; Wikstrom, Westermarck, and Palo 1976). However, subsequent studies indicate that the reduced glutathione peroxidase activity found in MS patients is independent of the selenium concentration and probably is due more to genetic factors.

While selenium supplementation may not increase the activity of glutathione peroxidase in the majority of patients with MS, it is a relatively inexpensive supplement that may benefit some patients. Antioxidant supplementation, particularly vitamin E, is definitely indicated, due not only to the increased lipid peroxidation noted in MS but also to the increased consumption of essential fatty acids. People with MS should supplement their diet with selenium—200 mcg per day—and vitamin E (d-alpha-tocopherol)—600 IU per day.

References

Bates, D., et al. (1978). "Polyunsaturated fatty acids in treatment of acute remitting multiple sclerosis." *Br Med J* ii:1390–1391.

Cunane, S. C. (ed.) (1991). Symposium Proceedings: Third Toronto Essential Fatty Acid Workshop on Alpha-Linolenic Acid in Human Nutrition and Disease. May 17-18, 1991, University of Toronto, Toronto, Ontario, Canada. *Nutrition* 7:435–446.

Horrobin, D. F. (1979). "Multiple sclerosis: The rational basis for treatment with colchicine and evening primrose oil." *Med Hypothesis* 5:365–378.

Jensen, G. E., Gissel-Nielsen, G., and Clausen, J. (1980). "Leukocyte glutathione peroxidase activity and selenium level in multiple sclerosis." *J Neurol Sci* 48:61–67.

Mazzella, G. L., et al. (1983). "Blood cells glutathione peroxidase activity and selenium in multiple sclerosis." *Eur Neurol* 22:442–446.

Millar, Z. H. D., et al. (1973). "Double-blind trial of linolate supplementation of the diet in multiple sclerosis." *Br Med J* i:765–768.

Paty, D. W., et al. (1978). "Linoleic acid in multiple sclerosis: Failure to show any therapeutic benefit." *Acta Neurol Scand* 58:53–58.

Shukla, V. K. S., Jensen, G. E., and Clausen, J. (1977). "Erythrocyte glutathione peroxidase deficiency in multiple sclerosis." *Acta Neurol Scand* 56:542–550.

Simopoulos, A. P. (1991). "Omega-3 fatty acids in health and disease and in growth and development." *Am J Clin Nutr* 54:438–463.

Swank, R. L. (1991). "Multiple sclerosis: Fat-oil relationship." *Nutrition* 7:368–376.

Swank, R. L., and Pullen, M. H. (1977). *The Multiple Sclerosis Diet Book*. Doubleday, Garden City, NY.

Swank, R. L., et al. (1952). "Multiple sclerosis in rural Norway: Its geographic distribution and occupational incidence in relation to nutrition." *New Engl J Med* 246:721–728.

Szeinberg, A., et al. (1979). "Decreased erythrocyte glutathione peroxidase activity in multiple sclerosis." *Acta Neurol Scand* 60:265–271.

Wikstrom, J., Westermarck, T., and Palo, J. (1976). "Selenium, vitamin E and copper in multiple sclerosis." *Acta Neurol Scand* 54:287–290.

FLAX AND CANCER

7

Dr. Johanna Budwig, considered Germany's premier biochemist and one of the world's leading experts on essential fatty acid nutrition, has built an international reputation for treating cancer and other degenerative diseases with flaxseed oil and a protein-rich diet. In 1990 the National Cancer Institute (NCI) launched a five-year, $20 million program to learn more about biologically active plant chemicals, phytochemicals, in certain foods that may help to prevent cancer. Flaxseed was the first of six foods to be studied. Preliminary results indicate that flaxseed oil can exert powerful anticancer properties if the oil is high in lignan precursors. Unfortunately, despite the incredible promise of preliminary results, the NCI project was canceled before it could be completed. Nevertheless, there is a substantial body of evidence that indicates that flaxseed oil exerts significant anticancer properties.

Flax Lignans: Potent Anticancer Compounds

In addition to their high level of omega-3 fatty acid, flaxseeds are also the most abundant source of lignans. One study shows flaxseeds contain over 100 times the levels found in other plant foods (Thompson et al. 1991). Lignans are special compounds that are demonstrating some rather impressive health benefits, including positive effects in relieving menopausal hot flashes, as well as anticancer, antibacterial, antifungal, and antiviral activity. The most significant of these actions of lignans are their anticancer effects (Lampe et al. 1994; Setchell and Adlercreutz 1988).

In animal experiments, tremendous anticancer effects have been noted when the animals are fed flaxseed and flaxseed oil. Positive effects have been shown not only in mammary cancer but also in colon cancer and general tumor models. Typically, the animals receiving flaxseed or flaxseed oil demonstrate significant reduction (for example, greater than 50 percent reduction) in tumor numbers and size after one to two months (Serraino and Thompson 1991, 1992a, 1992b).

Plant lignans are changed by the gut flora into enterolactone and enterodiol, two compounds believed to be protective against cancer, particularly breast cancer. Lignans are capable of binding to estrogen receptors and interfering with the cancer-promoting effects of estrogen on breast tissue. In addition, lignans increase the production of a special sex hormone–binding compound. This compound, known as *sex hormone–binding globulin,* regulates estrogen levels by escorting excess estrogen from the body via eliminative pathways.

Lignans are thought to be one of the protective factors against breast cancer in vegetarian women (Adlercreutz et al. 1986). Typically, women who excrete higher amounts of lignans in their urine (a sign of increased consumption) have much lower rates for breast cancer. High-lignan flaxseed oil may be the best choice for women going through menopause or women at risk for breast cancer. It is currently estimated that as many as one in nine American women will develop breast cancer.

Alpha-Linolenic Acid and Breast Cancer

In addition to lignans, a high-lignan flaxseed oil also provides alpha-linolenic acid. Like lignans, alpha-linolenic acid is also demonstrating significant anticancer properties, especially against breast cancer. A prospective study in 121 women with initially localized breast cancer examined the association between the levels of various fatty acids in the fatty tissue of the breast and how much the cancer had spread (metastasized). Breast tissue analyzed at the time of surgery showing a low level of alpha-linoleic acid (18:3w3) was associated with the spread of the cancer into the lymph nodes of the armpit (axillary area) as well as tumor invasiveness. After thirty-one months of follow-up after the initial surgery, twenty-one patients developed metastases of their cancer into other body tissues. Low levels of alpha-linolenic acid was the first determinant of metastases in these patients. In other words, when all factors were considered, low levels of alpha-linolenic acid was found to be the most significant contributor to the spread of cancer. Since the main cause of death in breast cancer patients is the development of cancer in other tissues, the significance of this finding is of extreme importance (Bougnoix et al. 1994).

The results from this study suggest that supplementing the diet with flaxseed oil (approximately 58 percent alpha-linolenic acid) may help prevent breast cancer, tumor invasiveness, and metastasis. In addition to this prospective human study, animal models of mammary carcinogenesis have shown that diets high in omega-6 fatty acids stimulate mammary tumor growth while alpha-linolenic acid enrichment of the diet inhibits tumor growth (Rose and Hatala 1994).

Modulating Effects of Fatty Acids on the Immune System

It is possible that alpha-linolenic acid also exerts some of its anticancer effects via enhancement of immune function (Kelley, 1992). In one study, the possible interaction between intense exercise, known to suppress the immune system, and polyunsaturated fatty acids was examined in mice (Benquet et al. 1994). For eight weeks the animals received either a natural ingredient diet only or a diet supplemented with 10 g/100 mg of flaxseed oil (50 percent alpha-linolenic acid), beef tallow, safflower oil (mostly linoleic acid), and fish oils. Each dietary group was divided into either a sedentary group or an exercised group. Exercise consisted of continuous swimming at high intensity until exhaustion. It was shown by three separate experiments that the primary immune response to sheep red blood cells was affected by supplementation with polyunsaturated fatty acids in sedentary animals in this order: beef tallow > control diet > safflower oil > fish oil > flaxseed oil. In the exercised animals, the immune response was suppressed by exhaustive exercise, except for the group receiving the flaxseed oil. Only the group receiving the flaxseed oil demonstrated a normal immune response.

The significance of these results to humans may be to recommend flaxseed oil supplementation to people who place great physical demands on themselves due to work or exercise to offset some of the negative effects of exercise on immune function.

References

Adlercreutz, H., et al. (1986). "Determination of urinary lignans and phytoestrogen metabolites, potential antiestrogens and anticarcinogens, in urine of women on various habitual diets." *J Steroid Biochem* 25:791–797.

Benquet, C., et al. (1994). "Modulation of exercise-induced immunosuppression by dietary polyunsaturated fatty acids in mice." *J Toxicol Environ Health* 43:225–237.

Bougnoix, P., et al. (1994). "Alpha-linolenic acid content of adipose breast tissue: A host determinant of the risk of early metastasis in breast cancer." *Br J Cancer* 70:330–334.

Kelley, D. S. (1992). "Alpha-linolenic acid and immune response." *Nutrition* 8:215–217.

Lampe, J. W., et al. (1994). "Urinary lignan and isoflavonoid excretion in premenopausal women consuming flaxseed powder." *Am J Clin Nutr* 60:122–128.

Rose, D. P., and Hatala, M. A. (1994). "Dietary fatty acids and breast cancer invasion and metastasis." *Nutr Cancer* 21:103–111.

Serraino, M., and Thompson, L. U. (1991). "The effect of flaxseed supplementation on early risk markers for mammary carcinogenesis." *Cancer Letters* 60:135–142.

Serraino M., and Thompson, L. U. (1992a). "Flaxseed supplementation and early markers of colon carcinogenesis." *Cancer Letters* 63:159–165.

Serraino, M., and Thompson, L. U. (1992b). "The effect of flaxseed supplementation on the initiation and promotional stages of mammary tumorigenesis." *Nutr Cancer* 17:153–159.

Setchell, K. D. R., and Adlercreutz, H. (1988). "Mammalian lignans and phytoestrogens: Recent studies on their formation, metabolism, and biological role in health and disease," in *Role of Gut Flora in Toxicology and Cancer,* I. R. Rowland (ed.). Academic Press, London, UK, 315–343.

Thompson, L. U., et al. (1991). "Mammalian lignan production from various foods." *Nutr Cancer* 16:43–52.

HOW TO SELECT A HIGH-QUALITY FLAXSEED OIL

8

Obtaining high-quality flaxseed oil and other healthful vegetable oils is a major challenge for consumers. In order to receive the benefits described within these pages, it is critical that a high-quality preparation be used. How can you be ensured that the product you pick off your grocer's shelf or buy in a health food store has been processed and handled in a way to protect the delicate polyunsaturated bonds these oils possess? Your purchase could bring you a product that truly supports life or one that has the potential to cause cellular destruction, immune malfunction, and ultimately, if used regularly, promote cardiovascular disease and cancer.

This paradoxical effect may be confusing unless you appreciate the characteristics of highly polyunsaturated vegetable oils like flaxseed oil. The analogy of Dr. Jekyll and Mr. Hyde is appropriate. Without ingesting his laboratory-derived potion, Dr. Jekyll is a well-intended scientist. However once the potion is ingested, Dr. Jekyll is transformed into the dangerous and evil Mr. Hyde. Such is the case with highly polyunsaturated oils. In their unrefined, unadulterated state, these nutrients are absolutely essential for life and optimal health; however, once refined or altered, these formerly life-imparting nutrients become transformed into potentially biologically dangerous compounds. Foods, including oils, also have a natural shelf life, unless they have been artificially preserved to increase their longevity.

People want foods that are fresh and oils that are not rancid. As a testament to the demand for foods that are fresh, it was found as a result of market surveys that the single word that most powerfully motivated consumers to buy was the word "fresh" on the label. Because highly polyunsaturated vegetable oils are considered semiperishable, the only way to extend shelf life is to artificially stabilize them to the point that they will not interact with the elements and thus become rancid. The greatest potential for destruction of polyunsaturated oils are heat, light, and oxygen.

Unfortunately, refined polyunsaturated vegetable oils are purposely exposed to temperatures as great as 700°F without protection from light and oxygen. The end results are the vegetable oils and artificial margarine spreads you find at your local grocery store, which is your first hint at where not to look for high-quality vegetable oils. The only retail outlets to date to address the important issues surrounding polyunsaturated vegetable oils are health food stores. Consumers, beware, because many of the therapeutic and culinary oils found in health food stores are also manufactured, distributed, packaged, and displayed in a fashion that seriously questions any health benefit they may have otherwise had.

To truly be assured of purchasing high-quality flaxseed oil and other essential fatty acid–rich oils, a further knowledge of extraction, handling, and labeling methods is required. First and most important to consider are what elements lead to the destruction and breakdown of vegetable oils, including flaxseed oil. In addition, the removal of naturally occurring antioxidant, vitamin, mineral, and other nutrient cofactors from the oil serves to accelerate the degradation of these products.

You would think that in the production of polyunsaturated oils, every possible way to avoid destructive manufacturing methods would be employed. Right? Wrong! In fact, this is the exception rather than the rule, as the majority of food-grade oil manufacturers purposefully expose polyunsaturated oils to excessive heat, light, and oxygen. Their goal is to achieve a uniform product with an exceptionally long shelf life, in the most cost-effective manner possible. Instead of these means accelerating the rancidity of the oil, the food manufacturer merely relies on the extremely high heat generated during the process (in excess of 250°C) to unnaturally contort the fatty acid molecule while stabilizing the unsaturated bond with hydrogen, yielding the extended shelf life they long for. Starting to sound like something Dr. Jekyll might have concocted in his laboratory?

Instead of going into a long and boring description of mass refinement of polyunsaturated oils, it is enough to say there are over twenty steps in this process that lead to the complete destruction of the health potential of these products. They are destroyed to the point where the manufacturer has to add vitamin E and beta-carotene again just to give the oil a tint of color. Perhaps we could refer to these oils as "enriched." Would you feel enriched if you were robbed at gun point, asked to strip naked, and then given your shoes back so you could walk home?

Avoiding Destructive Manufacturing Practices of Polyunsaturated Oils

Several methods of manufacturing have been recognized as destructive to polyunsaturated oils and the nutrient cofactors they contain (Anderson 1962; Beasley and Swift 1989). Avoid products produced in these ways if at all possible. Unfortunately, no federal laws have been established to enforce the listing of extraction and processing methods on the labels of vegetable oils. Needless to say, companies offering products extracted in these ways do not announce this fact on their labels. Your best bet in obtaining high-quality flaxseed and vegetable oils is by recognizing the hallmarks of these products. This information is given later in this chapter.

1. Mass commercialized oil products: These products are usually extracted in a number of ways using various combinations of extraction methods. For example, although these oils may initially be extracted by the preferred expeller press method, large commercial presses are utilized under extreme pressing temperatures and pressures in order to obtain high oil yields. The seed meal is then placed in a chemical solution in order to obtain every trace of oil, ultimately yielding 100 percent seed oil recovery. Then comes the remaining twenty or so refinement and stabilization steps. Hydrogenation is but one step that might be employed during the processing of oil products. Many of the negative consequences of hydrogenation have been previously discussed. An explanation of the procedure bears repeating. The process of hydrogenation effectively transforms an unsaturated fatty acid molecule into a fat similar to a saturated fat, as well as distorts the molecule from its healthful "cis" configuration to an unnatural, unhealthful "trans" configuration. Trans fatty acids occur when either the random process of hydrogenation or high heat straightens the formerly C-shaped polyunsaturated molecule. Though it would be difficult to get a bunch of C-shaped molecules to clump together, once they have been straightened, they clump together readily, solidifying the previously liquid oil as well as significantly raising the melting temperature. This process is what occurs in the manufacturing of margarine. Imagine margarine clumping together in your arteries and clogging up cell respiration! The simple yet effective process of hydrogenation fulfills all the requirements set forth by the food industry by reducing the likelihood of spoilage (now acts as a food preservative), extending shelf life, and enhancing palatability. However, the detrimental effects of hydrogenated oils are now well known (Enig 1993).

2. Hexane, or chemical extraction: Some seeds, because of their extremely small size, hardness, or low oil content, lend themselves to be extracted more efficiently and economically by chemical means. The seed from the evening primrose flower, a fair source of gamma-linoleic acid (GLA), is a good example. In hexane, or chemical extraction, the desired seed is first cracked to expose the oil, perhaps even run through an expeller press, and then immersed in a solvent solution to render the oil. The seed oil must then be separated from the solvent solution. There is some question as to whether all of the solvent solution can be removed from the resultant oil. We would have to ask from a consumer's point of view why one would purchase oils extracted in this fashion. While in its singular application hexane is a better extraction method than mass refined oils and supercritical fluid extracted oils, preferred seed oil extraction methods exist.

3. Supercritical fluid extracted oils: Supercritical fluid extraction (SCFE) is the latest player in seed oil extraction methods. This method employs an extremely highly pressurized chamber, similar to a pressure cooker, to remove the oil from the seed. Similar to hexane extraction, the seed is ground to expose more surface area for oil extraction. While highly touted by some companies marketing so-called nutritional oils, SCFE is even more damaging to polyunsaturated oils than hexane extraction. SCFE oils have less chemical stability, higher average lipid peroxide levels (measure of rancidity), altered fatty acid profiles, decreased mineral content, fractionated triglyceride formations, and are absent of phosphatides and stripped of tocopherols (Fabio, Mazzanti, and King 1991; King, Geary, and List 1990; List, Friedrich, and King 1989a, 1989b). Uninformed product buyers for nutrition companies are enamored with SCFE because low temperatures are employed during the extraction process. Unfortunately, SCFE, used singularly and with the exception of mass refined oils, renders the poorest quality oils available.

4. Cold pressed oils: Rarely in the food manufacturing business has there been a term so distorted as this one. "Cold pressed" by its very name conjures up oils being extracted while under conditions of refrigeration or other heat-free environments. Unfortunately in the United States, the only official designation given for cold pressed oils is that "no form of external heat is applied during extraction." This definition says nothing of the possible temperatures reached from the operation of the extraction apparatus or heat that the oils might be subjected to during further refinement after extraction. Needless to say, the term "cold pressed" displayed on a bottle of oil says absolutely nothing about the quality of the product. While this term has been well received by consumers, bringing about a false sense of security, quality manufacturers of unrefined vegetable oils have decided to drop this term from their labels because of the deception that surrounds it. Interestingly, in

Europe the term "cold pressed" is also used, but they have established regulations that allow the term to be used only if extraction temperatures are below 104°F. Similar regulations should be established in the United States if "cold pressed" is to be allowed for labeling purposes.

Preferred Oil Extraction Methods

Fortunately there are a few manufacturers of flaxseed oil and other polyunsaturated vegetable and seed oils that have realized the extreme need for unrefined oil products. These health-conscious companies are in the process of attempting to form a trade group that would establish regulations and commonly accepted terminology to aid consumers in distinguishing between "safe" and healthful nutritional and culinary oils and refined and "dangerous" fat and oil products that predominantly line our grocery and health food store shelves. The interested parties continue to meet and correspond in attempting to achieve this monumental task. Until such a time when industry regulations are established and common terminology is available, health-conscious consumers can use the information here to ensure purchasing quality oil products.

1. ***Expeller pressing:*** Expeller presses resemble a screw/nut mechanism where the seed is triturated between the threads and the body of the press head. The resulting pressure squeezes the oil from the seed. All manufacturers of quality oil products use some form of expeller press. Usually these manufacturers modify their expeller presses in order to reduce operating temperatures as well as pressure exerted on the seed. Unfortunately, expeller presses are also utilized by mass producers of oils. They use huge commercial presses that are operated at extreme temperatures and pressures for extended periods. In these instances, expeller presses are utilized only as a prerefinement step in the production of inferior oil products. While a significant improvement over other extraction methods, expeller presses must be used singularly and under stringent conditions to truly yield healthful oils.

2. ***Modified atmospheric packing (MAP):*** Modified atmospheric packing (MAP) is the term that is appearing to draw favor from manufacturers of high-quality oil products. This method employs the use of an expeller press, modified by the oil producer to operate at low temperatures while excluding the damaging effects of light and oxygen. Regulations to fully define the parameters at which a manufacturer would be privileged to use this designation are currently being discussed by the newly formed culinary and nutritional oils trade group. It may be some time before you will find this terminology designating these products on labels. Oil manufacturers currently utilize trade names for

their proprietary MAP methods, including Bio-Electron Process (Barlean's Organic Oils), Spectra-Vac (Spectrum Naturals), and Omegaflo (Omega Nutrition).

Table A
COMPARISON OF OILS EXTRACTED BY VARIOUS METHODS

	Modified Atmospheric Packing*	Super Critical Fluid Extraction (SCFE)	Hexane (Chemical Extraction)
Oxidative Stability	Highly Stable For Polyunsaturated	Highly Unstable	Poor Stability
Average Lipid Peroxide Level	0.1–0.3	1.7	1.3
Essential Fatty Acid Profile	Normal	Altered	Normal
Mineral Content	Normal	Decreased	Normal
Triglyceride Formation	Normal	Fractionated	Normal
Phosphotides	Present, Remain Intact	Absent (Refined Out)	Present, Remain Intact
Exposure to Light and Oxygen	Protected from Light and Oxygen	Exposed to Light and Oxygen	Exposed to Light and Oxygen
Chemical Residue	No	No	Yes
Color	Natural Pigment Intact	Pigments Removed During Process	Natural Pigment Intact
Taste	Smooth, Nutty Flavor	Bland, Bitter Aftertaste	Bland, Bitter Aftertaste
Shelf Life	Liquid–4 mos. (1 year in freezer) Capsules–1 year	Questionable as indicated by lipid peroxide level	Questionable as indicated by lipid peroxide level
Commercially Available Form	Liquid or Capsule	Capsule Only	Liquid or Capsule

* Example utilizing Barlean's Organic Oils "Bio-Electron" MAP process. Results may vary with other MAP extraction processes.

Hallmarks of Quality Oil Products

Luckily there are ways in which educated health-conscious consumers can be ensured of selecting high-quality flaxseed and other vegetable and seed oils. Simply look for these hallmarks:

- ***Products certified as organic by a reputable third-party source, indicated on label or promotional material.*** Manufacturers that have devoted the time and expense to have their oil products third-party certified organic have almost always taken the measures necessary to provide extremely high-quality oils. They will proudly proclaim this status on their label.

- ***Products extracted by modified expeller presses only and at temperatures that do not exceed 98°F.*** Again, manufacturers interested in human health will not sacrifice quality for the higher oil yields extreme temperatures and pressures bring. They use methods that exclude light and oxygen during manufacture. These methods are currently recognized by several popular trade names listed above under "modified atmospheric packing."
- ***Products contained in opaque (light-resistant) plastic containers.*** Total elimination of light coming in contact with the oil is mandatory to ensure against degradation of the oil. Every moment that the oil is exposed to light, potentially thousands of photons interact with the oil, ultimately causing a rancid product. The light transmission of the various light frequencies combined results in a mean penetration of even pharmaceutical amber glass bottles of 114 percent. This percentage is in stark contrast to the 0 percent penetration allowed by opaque plastic bottles. Multiply 114 percent times the number of hours and days the oil is exposed to light and you get a good idea of the potential impact on the oil. Only purchase flaxseed oil that is contained in totally opaque, high-density polyethylene (HDPE) plastic bottles. Despite the accusations of an author on fats and oils, who is secured by a company that only packages in glass, the use of high-density polyethylene plastic is absolutely safe and the preferred method of packaging these oils. HDPE plastic is fully approved by the U.S. and Canadian governments for these purposes. This material has been used since the 1970s with an untarnished record of health and safety. Extensive laboratory analysis has been conducted to ensure absolutely no migration of HDPE plastic materials into oils and other foods. No scientific evidence exists to date to refute the extreme safety of this packaging material.
- ***Products delivered manufacturer-direct to retail health food stores or your home.*** Manufacturer-direct delivery avoids any prolonged layover these products may have while sitting in a distributor warehouse.
- ***Products recommended by reputable health and nutrition authorities.*** This may include paid or nonpaid endorsements. Rarely will a well-respected authority stake his or her reputation on an inferior product. Just because a product does not carry an endorsement doesn't mean it is not a quality product.

- ***Products found in the refrigerated section of health food stores.*** Manufacturers of high-quality, unrefined oil products insist that their products be stored in refrigerators to protect the oil from degradation from prolonged exposure to ambient temperatures.
- ***Manufacturers that are willing to supply you with a third-party laboratory analysis of their product upon your request.*** Manufacturers of high-quality oil products are happy to supply you with these laboratory analyses attesting to the superior quality of their oils.
- ***Products backed with educationally based supportive materials.*** Manufacturers of high-quality oil products typically disseminate educational materials to health food stores and individuals.
- ***Quality vegetable oil products will have dating codes on the label signifying the pressing date and the recommended use-by date.*** Follow these dates stringently, unless you decide to place your flaxseed oil in the freezer, in which case you can extend the expiration date significantly.

Our personal recommendations to simplify your search is to ask for Barlean's Organic Flax Oil, as we believe that they manufacture the finest flaxseed oil anywhere in the world. Apparently many of the world's top authorities on health and nutrition agree with us, including the world's leading authority on the topic of fats and oils, German biochemist Dr. Johanna Budwig, who recommends only Barlean's Organic Flax Oil. Women's health expert and author Ann Louise Gittleman, considered one of the top-ten nutritionists in the nation, endorses Barlean's Organic Oils. Dr. Robert Erdmann, recognized as an international figure in the field of nutritional biochemistry as well as a pioneer of amino acid therapy, also uses and recommends only Barlean's for his personal and professional use.

Barlean's began as a small family company. Owner Bruce Barlean and his father, David Barlean, invented a method to extract flax oil from the seeds to protect the oil at every step of the manufacturing process—the result, an oil that simply cannot be matched in taste and quality. Barlean's acclaim quickly grew as word of mouth spread the extreme commitment of the Barleans and the quality of their products. Today, Barlean's is no longer a "best kept secret" as their products are placed in over 2,000 health food stores nationwide as well as countless health professionals' offices. Despite the recognition that Barlean's has received, the Barleans and their company are as down-to-earth and friendly as the day they opened their doors. Most important, they refuse to adopt any cost-cutting measures that would compromise their product. They have learned to rely on what they know and do best: press organic vegetable oils. For questions, comments, or to receive educational materials regarding flaxseed oil, write or call:

Barlean's Organic Oils
4936 Lake Terrell Road
Ferndale, WA 98248
(800) 445-3529

Avoid These Oil Products

- Oils that are not third-party certified organic
- Oils that are sold only in gelatin capsule form
- Products where the actual method of extraction is not listed on the label. Be wary of "cold pressed" oils
- Oils that are heavily refined or that have a bitter aftertaste, rancid flavor, or no flavor at all

Summing Up

To truly realize the awesome potential flaxseed oil has on human health, you must first choose a product that has been manufactured and handled under very stringent conditions in order to protect the oil from degradation. Using the guidelines presented in this chapter, you will be ensured of choosing oil products that truly support life in every way.

References

Anderson, A. C. (1962). *Refining of Oils and Fats.* Pergamon Press, New York.

Beasley, J. D., and Swift, J. J. (1989). *The Kellogg Report: The Impact of Nutrition, Environment & Lifestyle on the Health of Americans.* Institute of Health Policy and Practice, Bard College Center, Annandale-on-Hudson, NY, 135–143.

Enig, M. G. (1993). "Trans fatty acids: An update." *Nutrition Quarterly* 17:79–95.

Fabio, F., Mazzanti, M., and King, J. (1991). "Supercritical carbon dioxide extraction of evening primrose oil." *JAOCS* 68:422–427.

King, J., Geary, B., and List, G. R. (July 1990). "A solution thermodynamic study of soybean oil/solvent systems by inverse gas chromatography." *JAOCS* 67:424–430.

List, G. R., Friedrich, J. P., and King, J. (1989a). "Oxidative stability of seed oils extracted with supercritical carbon dioxide." *JAOCS* 66:98–100.

List, G. R., Friedrich, J. P., and King, J. (1989b). "Supercritical CO_2 extraction and processing of oilseeds." *Oil Mill Gazette* Dec:28–34.

FLAXSEED OIL RECIPES*

9

There is no better way to enjoy the virtues of essential fatty acid–rich flaxseed oil than to include it in some of your favorite recipes. Because the vital and essential fatty acids are being very effectively removed or transformed into dangerous trans and hydrogenated fatty acids in our food chain by modern manufacturing methods, it is necessary for us to reintroduce the essential fatty acids back into our foods. Utilizing flaxseed oil in recipes is a terrific way to ensure that you get your daily requirement of essential fatty acids (Carter 1993; Stitt 1988). You should feel free to experiment with adding flaxseed oil to foods. Salad dressings render themselves especially suitable for the inclusion of flaxseed oil. Remember to consider the destruction that can come with excessively heating flaxseed oil. If at all possible, add flaxseed oil to already cooked foods like soups. With that in mind, enjoy the culinary delights and health benefits a diet rich in essential fatty acids can bring.

*Recipes compliments of Barlean's Organic Oils.

Barlean's Basic Salad Dressing

(MAKES 2–3 LARGE SALAD SERVINGS)

Prepare this dressing in your salad bowl. Place all ingredients in a salad bowl and whisk together until smooth and creamy. This is quick and delicious!

4 tablespoons Barlean's Organic Flax Oil
1 1/2 tablespoons lemon juice
1 medium clove garlic, crushed
pinch of seasoned salt or salt-free seasoning
freshly ground pepper to taste

Using your favorite spices, jazz up your salad to your own personal taste.

Barlean Butter

At last an easy-to-make, healthy bread spread to replace margarine. Tastes wonderful on top of fresh, steamed vegetables.

Use equal parts of butter and Barlean's Organic Flax Oil

Melt the butter in a saucepan over low heat. Pour the butter into a container and stir in the flax oil. Chill until hardened.

Pumpkin Seed–Mint Sauce

(MAKES ABOUT 1 1/2 CUPS)

Delicious over your favorite grains or vegetables.

2 tablespoons mayonnaise
1 cup hot water
1/4 cup pumpkin seeds, toasted lightly
2 large green serrano chilies, seeded and chopped
1/2 teaspoon onion powder
1/2 teaspoon salt or salt substitute
2 tablespoons dried mint or 3 tablespoons fresh mint
3 cloves garlic, crushed
3 tablespoons Barlean's Organic Flax Oil
black pepper to taste

Combine all ingredients in a blender or food processor, and process until creamy.

Pour over vegetables or grains.

Eggplant Dip

(MAKES ABOUT 2 CUPS)

Wonderful as a vegetable dip or sandwich spread.

1 eggplant
2 medium cloves garlic, crushed
1 green onion, chopped
1/4 cup chopped parsley
1 tablespoon lemon juice
3 tablespoons Barlean's Organic Flax Oil
1/2 teaspoon dill weed

Bake the eggplant at 400°F for 1 hour or until soft.

Remove from oven and when cool enough to handle, peel and dice.

Place all ingredients in a blender or food processor and process until smooth.

Chill and serve.

Hummus

(MAKES ABOUT 2 1/2 CUPS)

A fantastic-tasting Middle Eastern dish to be used as a dip or as a filling in pita pocket sandwiches. An excellent source of complete protein and, now, essential fatty acids.

1 15-ounce can or 1 2/3 cups cooked garbanzo beans (chickpeas)
1/4 cup tahini (sesame seed paste)
3 tablespoons lemon juice
3 tablespoons Barlean's Organic Flax Oil
2 medium cloves garlic
1/4 teaspoon ground coriander
1/4 teaspoon ground cumin
1/4 teaspoon paprika
dash of cayenne
1/4 cup minced scallions
2 tablespoons minced fresh parsley for garnish

In a blender or food processor, process the garbanzo beans, tahini, lemon juice, and flax oil until the mixture reaches the consistency of a coarse paste. Use as much of the garbanzo liquid, or water, as needed. Add the garlic, coriander, cumin, paprika, and cayenne and blend thoroughly.

Transfer the hummus to a bowl and stir in the scallions.

Cover the hummus and chill.

Garnish with parsley before serving.

Barlean's Tangy Dip

(MAKES ABOUT 1 CUP)

Great with vegetables, crackers, or spread on bread.

6 ounces plain yogurt
3 tablespoons Barlean's Organic Flax Oil
pinch of herbal or salt seasoning
1/2 teaspoon dill weed
1 tablespoon marjoram
2 medium cloves garlic, crushed

Combine and mix all ingredients. Chill for 1/2 hour.

Fresh Mexican Salsa

(MAKES 2 CUPS)

A zesty traditional Mexican salsa made even better with the addition of flaxseed oil. Great as a dip for tortilla chips or as a sauce on enchiladas, burritos, and tacos.

3 tomatoes, diced
4 sprigs fresh cilantro
1/2 medium onion, diced
1 green onion, chopped
1 small jalapeno pepper
1/2 cup tomato sauce
3 tablespoons Barlean's Organic Flax Oil

Combine the tomatoes, cilantro, onions, and jalapeno pepper in a blender or food processor and process to desired consistency, chunky or saucy.

In a separate bowl, combine the tomato sauce and flax oil. Stir to a uniform consistency.

Combine the mixtures and chill until ready to serve.

Happy Apple Breakfast

(SERVES 2)

Other rolled grains such as wheat, barley, rye, or triticale can be used in various combinations instead of some or all of the rolled oats. All of these grains have the same cooking characteristics.

1 1/2 cups rolled oats
2 1/2 cups water, milk, or apple juice
1 medium green or golden apple, sliced
1/4 cup currants, raisins, or chopped dates
1/4 cup freshly roasted pumpkin seeds (optional)
2 tablespoons Barlean's Organic Flax Oil
raw honey to sweeten
plain nonfat yogurt to top (optional)

Combine all ingredients (except the flax oil and honey) in a saucepan. Simmer on the stove for 10 minutes. Cover and let stand for 5 minutes.

Stir in the flax oil and honey to taste.

Top with yogurt if desired.

French Toast with Flax-Maple Syrup

(SERVES 2–3)

French toast with a buttery-like maple syrup made by adding flaxseed oil.

French Toast:

1/2 cup whole wheat pastry flour
1/4 teaspoon salt or salt substitute
1 cup rice milk or soy milk
6 slices whole-grain bread
cinnamon and nutmeg to taste

Mix the flour and salt in a bowl large enough to accommodate one piece of toast. Aerate mixture with a wire whisk.

Pour the milk into the center of the flour and mix briskly.

Let batter stand 20 minutes.

Heat a lightly buttered griddle. Mix the batter well. Dip each slice of bread in batter to coat completely. Cook until the first side is lightly browned (about 3 minutes), then turn over and cook second side.

Sprinkle with cinnamon and nutmeg to taste.

Syrup:

1/4 cup pure maple syrup
1/4 cup Barlean's Organic Flax Oil

Whisk the maple syrup and flax oil until thoroughly blended.

Top French toast with flax-maple syrup.

Apple Muesli

(SERVES 2)

Popularized in Europe, muesli is a tremendously healthy start to your day.

2 tablespoons oatmeal
4 teaspoons water
2 apples
2 1/2 tablespoons wheat germ
juice of 1/2 lemon
3/4 cup yogurt
1 tablespoon raisins
2 tablespoons Barlean's Organic Flax Oil
2 tablespoons raw honey
3 tablespoons chopped walnuts

Soak the oatmeal overnight in the water.

Grate the apple or process in a food processor.

Combine all ingredients and mix well. Eat immediately.

Barlean's Quick and Easy Oatmeal

(SERVES 2–3)

A quick and simple version of Barlean's Best Oatmeal. A nutritious and healthy way to start your day.

1 cup quick oats
2 cups rice milk, soy milk, milk, or water
1/4 cup raisins
1/4 teaspoon vanilla
dash of cinnamon
raw honey to flavor
2 tablespoons Barlean's Organic Flax Oil
plain nonfat yogurt to top

Combine the oats, milk, and raisins in a saucepan. Bring to a boil. Cook about 1 minute over medium heat, stirring occasionally.

Remove from heat. Add the vanilla, cinnamon, and honey. Stir in the flax oil.

Serve topped with yogurt.

Barlean's Bio-Electron Smoothie

(MAKES 2–3 SERVINGS)

A delicious combination rich in vitamins, minerals, protein, fiber, and essential fatty acids for your good health.

Combine in a blender in descending order:

1/2 cup rice milk, soy milk, or low-fat milk
1 tablespoon Barlean's Organic Flax Oil
3 tablespoons (heaping) plain yogurt
1 orange, peeled and sliced
1/2 apple, sliced
1 frozen banana, sliced
4 ice cubes
protein powder (optional)

Add 1/2 to 3/4 cup of one or more of the following fruits (fresh or frozen) as a dominant flavor: strawberries, blueberries, blackberries, raspberries, peaches, pineapple.

Blend on high speed until smooth.

References

Carter, J. F. (1993). "Potential of flaxseed and flaxseed oil in baked goods and other products in human nutrition." *Cereal Foods World* 38:753–759.

Stitt, P. (1988). "Efficacy of feeding flax to humans and other animals." Proceedings, Flax Institute 52:37–40.

ANSWERS TO COMMON QUESTIONS ABOUT FLAXSEED OIL

APPENDIX A

Because of the almost unbelievable healing potential and the increased popularity of unrefined, organic flaxseed oil, numerous questions are often raised about this remarkable food. Many misconceptions can surround nutritional supplements and super foods (foods unique for their extraordinary nutrient content or incredible healing potential). This appendix is devoted to answering some of the most common questions raised about flaxseed oil.

What is flaxseed oil?

Flaxseed oil is classified as a polyunsaturated vegetable oil. While polyunsaturated vegetable oils are relatively common, most of these oils have been damaged by processing methods. Flaxseed oil is unique to polyunsaturated fatty acids because it is the richest source of both essential fatty acids.

What are essential fatty acids and why are they important?

Essential fatty acids are two of forty-nine known "essential nutrients" that must be consumed in foods and cannot be manufactured (biosynthesized) by the human body. The essential fatty acids found primarily in vegetable oils are the most susceptible to destruction caused by manufacturing methods. Much of the popularity surrounding unrefined and organic flaxseed oil is based on the fact that it supplies ample amounts of both the omega-3 and omega-6 fatty acids.

With all the hype surrounding a number of dietary supplements on the market, why should I consider adding flaxseed oil to my nutritional regimen?

Evidence now exists to suggest a widespread deficiency of the essential fatty acids that flaxseed oil supplies. Essential fatty acids are directly and indirectly responsible for and involved with countless important and life-sustaining biological functions. A lack of essential fatty acids in the diet has been associated with numerous diseases and health complications, including heart disease, strokes, and cancer. Unlike many commercially available foods and nutritional supplements, the essential fatty acids found most abundantly in flaxseed oil are truly "essential" to life.

Who can benefit from taking flaxseed oil?

Because of the virtual elimination of essential fatty acids from our food chain, and the fact that flaxseed oil contains both of the essential fatty acids required for optimal health, almost everyone can benefit from taking flaxseed oil.

How is flaxseed oil found in a health food store different from other less-expensive vegetable oils that I might find in my local grocery store?

Oil products found in your local grocery store have commonly been subjected to a very harsh refinement process that either eliminates the essential fatty acids or transforms them into toxic compounds. The majority of flaxseed oil products found in health food stores have been protected from harsh refinement and thus preserve the vital essential fatty acids they contain.

How can flaxseed oil have such profound effects with so many health problems?

This is because of the extreme lack of the essential fatty acids in our modern diet. Once the essentials are returned to the diet, the body is equipped to carry out the biological functions the essential fatty acids are prerequisite to performing. For this reason numerous diseases and health problems are improved or alleviated. To date, over sixty illnesses and health problems are related to fatty acid abnormalities.

Has flax been found to be effective in the treatment or prevention of cancer?

There have now been numerous studies conducted on flaxseed and flaxseed oil to attest to the therapeutic and preventive properties against cancer. Most notable are the ability of the lignans, contained in the shell matrix of the flaxseed, to prevent colon and breast cancer. Other studies conducted with flax have found as great as a 50 percent reduction of tumor size in animal studies. Significant groundbreaking work regarding flaxseed oil and cancer has been done by German biochemist and Nobel Prize nominee Dr. Johanna Budwig.

What are "lignans" and are they found in flaxseed oil?

Lignans are a class of highly researched plant chemicals (phytochemicals) that have been found to have anticancer, antifungal, antibacterial, and antiviral properties. Their greatest attribute is in the possible prevention of breast and colon cancers. The highest concentration of lignans, compared to any other food, is found in the hull of flaxseeds. Generally, flaxseed oil does not contain lignans because the fragmented hull settles out of the oil during processing. The exception to the rule is a "High in Lignan" flaxseed oil product marketed by Barlean's Organic Oils, where the fine flaxseed hull particulate has been retained in the oil.

How much flaxseed oil should one take on a daily basis?

For most people 1 tablespoon for every 100 pounds of body weight is the rule. Generally, individuals who are ill should begin by taking as little as half a tablespoon a day in order to develop a tolerance for essential fatty acid–rich oil. Within two to three weeks they should be able to easily tolerate a full tablespoon.

Can you take too much flaxseed oil?

All food, nutrients, and elements including water and oxygen, when consumed, ingested, or breathed at excessive levels become toxic to the human organism. With flaxseed oil a safe guideline is the recommendation in the preceding answers. However, flaxseed oil has been used in much larger dosages in the treatment of cancer and other diseases. In these instances it is strongly advised that you are under the guidance of a qualified naturopathic or holistic medical doctor.

Can I expect to get the same benefits from flaxseed oil found in gelatin capsules as I would from the straight liquid oil?

Generally, one should be wary of oil products sold in gelatin capsules. The only way to test encapsulated oils for quality is to puncture the capsule and taste the oil. If the oil has a rich, robust, and nutty flavor without a bitter aftertaste, chances are the oil is unrefined and relatively fresh. If there is an off-flavor followed by a stinging sensation, you have swallowed rancid oil. If there is little or no taste, you have just sampled a refined oil product. Oil products found in gelatin capsules are not as high quality because of the extra manufacturing step taken to encapsulate them. In addition, the pure liquid oil is far more economical, especially when you consider that you have to take at least nine 1000 mg flaxseed oil capsules just to get 1 tablespoon of flaxseed oil. On the other hand, for someone who would not otherwise take the oil or for convenience during traveling, capsules are the solution. Rely on only manufacturers of high-quality oil products when purchasing flaxseed oil capsules.

What is the preferred material for packaging flaxseed oil—plastic or glass?

An opaque plastic container made of high-density polyethylene (HDPE) is the preferred material for packaging and protecting flaxseed oil from light. HDPE plastic is fully approved by the U.S. and Canadian governments for these purposes and has an untarnished record of health and safety. Independent laboratory analysis conducted by responsible organic oil producers have resulted in absolutely no migration of the HDPE plastic into the oil. Even amber pharmaceutical-grade glass allows over five different light frequencies to penetrate the bottle, potentially destroying the benefits of the oil.

Are there any side effects common to taking flaxseed oil?

Because flaxseed oil is simply a food source, side effects from supplementing with flaxseed oil are highly uncommon. The possibility does always exist however, just as with any food source, that someone may react unfavorably to the oil. For some individuals this may be a transitory effect where simply reducing the dosage should relieve any problem. For others, simply discontinue usage or seek the advice of a nutritionally oriented practitioner.

Is there a preference for taking flaxseed oil alone or with food?

There are several advantages to taking flaxseed oil with another food source. Mixing flaxseed oil with yogurt, for example, helps to emulsify the oil, aiding in optimal digestion, absorption, and utilization of the essential fatty acids. Based on the work of Dr. Budwig, adding flaxseed oil to foods rich in sulfated amino acids, such as vegetables of the cabbage family and cultured dairy products, helps in incorporating the essential fatty acids into cellular membranes.

Because flaxseed oil is a polyunsaturated vegetable oil and susceptible to damage from heat, does the transit time from the manufacturer to the health food store degrade the oil?

Either high heat or sustained heat over a long period of time can cause a degradation of unrefined flaxseed oil. The relatively short transit times and variable temperatures experienced in delivering flaxseed oil to health food stores has been found via testing to be insignificant. Part of the reason for this is because of the extremely low beginning levels of lipid peroxides (measure of rancidity in oils) in unrefined flaxseed oil products. Despite the unsaturated bonds flaxseed oil possesses, unrefined products are much more stable than most people think.

Why is it important to refrigerate flaxseed oil?

Although limited exposure to room temperatures has virtually no effect on flaxseed oil, prolonged exposure can begin to break down the product, eventually leading to rancidity. Better than refrigeration, consider placing your flaxseed oil in the freezer. Because of the many polyunsaturated bonds flaxseed oil possesses, the product will not usually freeze unless you have an extremely cold freezer.

Why should I consider using flaxseed oil over fish oil supplements?

A significant body of evidence exists to attest to the numerous therapeutic benefits of fish oils. Unfortunately, there also some hazards associated with fish oil supplementation. Encapsulated fish oil products have been associated with having extremely high levels of lipid peroxides that at least one study equated as a possible contributor to some forms of cancer. The high level of lipid peroxides found in fish oils also requires that people take extra measures in the form of antioxidant nutrients to protect themselves against free-radical destruction. Studies conducted on unrefined flaxseed oil has not exhibited the same antioxidant wasting effect. However, supplementing your diet with additional antioxidant protection is always a good idea when taking any highly polyunsaturated fatty acid.

Is flaxseed oil expensive, and where can I find it?

Flaxseed oil is the most health-enhancing oil product on the market, yet bears a significant price advantage when compared to other nutritional oil products. However, it must be pointed out that unrefined flaxseed oil is more expensive than refined oil products found at your grocery store. You can generally expect to pay between $6.50 and $8.00 for an 8-ounce bottle. Most health food stores keep high-quality flaxseed oil products stocked either in a refrigerator in their supplement department or near their dairy products. Flaxseed oil can be obtained through mail order with some companies, and at least one reputable network marketing company offers high-quality flaxseed oil.

ESSENTIAL FATTY ACIDS AND HEALTH CONDITIONS

APPENDIX B

Health Conditions Exhibiting Fatty Acid Deficiencies and Improvement with Appropriate Fatty Acid Supplementation

Acne
AIDS
Allergies
Alzheimer's
Angina
Angioplasty
Atherosclerosis
Arthritis
Autoimmunity
Behavioral Disorders
High Blood Pressure
Breast Cancer
Breast Cysts
Breast Pain
Cancer
Cartilage Destruction
Coronary Bypass
Cystic Fibrosis
Dementia
Dermatitis
Diabetes
Eczema
E. Coli Infection
Heart Disease
Hyperactivity
Hypertension
Hypoxia
Ichthyosis
Immune Disorders
Infant Nutrition
Inflammatory Conditions
Intestinal Disorders
Kidney Function
Learning
Leprosy
Leukemia
Lupus
Malnutrition
Mastaglia
Menopause
Mental Illness
Metastasis
Multiple Sclerosis
Myocardial Infarction
Myopathy
Neurological Disease
Obesity
Osteoarthritis
Post Viral Fatigue
Pregnancy
Psoriasis
Refsums Syndrome
Reyes Syndrome
Rheumatoid Arthritis
Schizophrenia
Sepsis
Sjogren-Larson Syndrome
Stroke
Vascular Disease
Vision

RESOURCES AND RECOMMENDED READING

To locate the services of a naturopathic physician in your area write or call:

The American Association of Naturopathic Physicians
P.O. Box 20386
Seattle, WA 98102
(206) 323-7610

Additional Recommended Reading

Fats That Can Save Your Life
Author: Dr. Robert Erdmann

A complete and entertaining sourcebook outlining the power of the essential fatty acids found in flax and the dramatic and even lifesaving capabilities they possess.

Flax Oil as a True Aid Against Arthritis, Heart Infarction, Cancer and Other Diseases
Author: Dr. Johanna Budwig

A comprehensive review of the groundbreaking research of the world's leading expert on fats and oils nutrition and flaxseed oil, German biochemist and Nobel Prize nominee Dr. Johanna Budwig. An in-depth study of the full potential of flax in the human diet.

These two books are available through Barlean's Organic Oils by calling or writing:

Barlean's Organic Oils
4936 Lake Terrell Road
Ferndale, WA 98248
(800) 445-3529

The revised *Beyond Pritikin*

Author: Ann Louise Gittleman

Chronicles Miss Gittleman's experience with the Pritikin longevity centers and her decision to employ essential fatty acids in her clinical practice. The new revision of the original 1988 edition contains all the latest information on the healing essential fatty acids. An extremely enjoyable read for anyone interested in health and essential fatty acids.

Available through Uni-Key by calling or writing:

Uni-Key
P.O. Box 7168
Bozeman, Montana 59771
(800) 888-4353